Urban Planning Theory Since 1945

Nigel Taylor

SAGE Publications
London • Thousand Oaks • New Delhi

To my father, with admiration and love,
and all those others who endured or died in
the Second World War hoping for a better future

First published 1998

SAGE Publications Ltd
6 Bonhill Street
London EC2A 4PU

SAGE Publications Inc
2455 Teller Road
Thousand Oaks, California 91320

SAGE Publications India Pvt Ltd
32, M-Block Market
Greater Kailash – I
New Delhi 110 048

British Library Cataloguing in Publication data

A catalogue record for this book is
available from the British Library

ISBN 0 7619 6094 5
ISBN 0 7619 6093 7 (pbk)

Library of Congress catalog card number record available

Typeset by Dorwyn Ltd, Rowlands Castle, Hants
Printed in Great Britain by Athenaeum Press, Gateshead

Contents

Acknowledgements iv

Introduction v

Part I Early post-war planning theory

1 Town planning as physical planning and design 3

2 The values of post-war planning theory 20

3 Early critiques of post-war planning theory 38

Part II Planning theory in the 1960s

4 The systems and rational process views of planning 59

5 Planning as a political process 75

Part III Planning theory from the 1970s to the 1990s

6 Theory about the effects of planning 95

7 Rational planning and implementation 111

8 Planning theory after the New Right 130

Part IV Conclusion

9 Paradigm shifts, modernism and postmodernism 157

Bibliography and references 170

Index 181

Acknowledgements

Hugh Barton, Martin Chick, Ron Griffiths and Tony Scrase were kind enough to read earlier drafts of selected chapters of this book. I am very grateful to them for their comments, especially as, for the most part, I came to agree with their critical comments and modified the text accordingly. However, I of course am responsible for what I have written.

I also thank my wife, Diana, and my children Georgia and Lawrence, for all their support and love. Without them I don't know if I would have bothered with this. But having decided to bother, I often stayed on in my office at work to do the task, and so was often a late-home husband and father. Never once was there any complaint or resentment. On the contrary, always a welcome. The loss was as much my own; I'd have much rather been home. But for some reason some of us are compelled to do these things, and I hope that the product will be worth it, for my students if not for my family.

Introduction

This book describes the history of town planning theory since the end of the Second World War (1945). Over this fifty-year period ideas about town planning have changed significantly. Yet students of town planning lack a book which describes, in an accessible way, the recent development of ideas which have informed their discipline. This book aims to fulfil that purpose.

As part of their town planning studies, students usually take some course in 'planning theory'. But as I know from my own experience of teaching this subject, students find the subject difficult. Part of this difficulty may be due to the intrinsic nature of the subject-matter, which deals with ideas and arguments rather than the accumulation and transmission of facts about planning. But the difficulties which students experience are not eased by the literature of planning theory. Much of the original literature in the subject is unnecessarily complicated and obscure, and so pretty impenetrable to the average student. Enthusiasm kindled in the opening week of a course on the subject can soon be drowned by the first reading of some 'classic text' in planning theory! There are some useful 'readers' in planning theory, such as Andreas Faludi's reader published in 1973 (Faludi, 1973a), and the more recent reader put together by Scott Campbell and Susan Fainstein (1996).[1] However, what is still lacking is a book which 'tells the story' of how town planning theory has changed since the end of the Second World War. Again, I have tried in this book to meet that lack. In so doing I have tried to tell the (his)story in a clear and accessible way, without sacrificing analytical rigour. For in my view, a book on the history of ideas should not only describe the ideas under consideration but also draw the reader into assessing them. Whether or not I have succeeded in these aims I leave for others to judge.

Before I begin I should say something about what I take 'town planning theory' to be (this itself has been a matter of debate amongst planning theorists since 1945). On this, it is worth saying to begin with that, if the practice of town planning is, literally, actually doing it, then everything that town planning students do at college is 'theory' about town planning of one kind or another, even when, for example, they are learning about the law that governs town planning. What is distinctive about the subject of 'planning theory' is that it aims to provide some overall or *general* understanding of the nature of town planning. Because of this, the sorts of questions planning theorists ask (or should ask) about town planning are *fundamental* questions about town planning. Questions such as: What sort of an activity is town planning? What should town planning be aiming to do? What are the effects of actual town planning practice? Because these are basic questions about town planning, they are also 'simple' questions; thcy are the sorts of questions a child might ask about town planning. But, as anyone who has had children will know, the 'simple' (i.e. fundamental) questions are generally the most difficult to answer, because they probe to the very basis of our thought.[2] This, then, is another reason why many students may find the subject of 'planning theory' difficult and why also, in spite of my best efforts to make this book straightforward and accessible, some parts of it may still make difficult reading.

A brief note, too, about the term 'town' planning. I use this term for simplicity's sake and also because it is probably the most widely used term the world over to describe the activity I am concerned with. But in using 'town planning' I take it to refer also to what some people (especially in the USA) call 'urban' or 'city' planning. All the terms 'town', 'urban' and 'city' make it clear that the focus of this discipline is the (planning of) the built environment. However, the way we fashion the urban (built) environment also affects the rural and natural environment, and so we should perhaps rename our activity 'environmental' planning (some texts on 'town' planning do carry this title). What I describe here as 'town' planning also encompasses what in Britain has been traditionally, and charmingly, called 'town and country' planning.

The story I have told is from a British perspective. But since the Second World War, the kind of town planning that has been practised and debated in Britain has been similar in many respects to town planning in other advanced capitalist democracies. So even though what I describe here is the development of town planning thought in Britain, much of this thought came from other places, especially the USA. I therefore hope that readers in other countries will find this account of interest and relevance to them.

So much for the terms of this text. More important than all this is the following fact: *in the twentieth century, most people in Europe and North America, and now increasingly in other parts of the world too, have come to live in cities*. This 'move to the city' has been associated with a great sense of loss for something which the countryside, or 'wild nature', provided, and in Britain this has generated a 'rural nostalgia' and a tradition of 'anti-urban' literature (see, e.g. Williams, 1973). This has played a significant part in twentieth-century town planning thought. I don't think we should belittle these sentiments, for it may be that, in the prescence of 'wild nature', many people experience something sublime and transcendent which is generally not available in cities. On the other hand, cities can be wonderful places, as is indicated by the numbers of people who flock to see cities like Florence and Venice, Paris and Rome, London and New York. But these, perhaps, are the exceptions. If there is another source of the anti-urbanism which has been so prevalent, it is that many cities are inhospitable, ugly places. But cities are human-made things, and the fact that some cities are congenial and uplifting shows that the miserable urban environments which most people are condemned to live in don't *have* to be like that.

It is this which makes town planning important. Indeed, it is an extraordinary fact that, in our contemporary urban culture, the activity of town planning is not more widely discussed and written about, and so, apparently, not perceived as being very significant in relation to the quality of our lives. The point has been well made by the novelist Margaret Drabble (1991, p. 32):

> I was recently talking with friends about which of the arts has the most powerful and direct effect upon the emotions. The rival claims of music and poetry found the most powerful advocates, until one unexpectedly nominated architecture. A surprised and respectful silence fell. Architecture? Did she really mean architecture? Did *buildings* make her want to weep or sing with joy? We questioned her and, yes, she did mean buildings . . .
>
> I have thought back to this discussion many times, and now consider this friend's point is less eccentric than at first appeared. Some of the greatest and grandest emotional and aesthetic experiences come from architecture. Who can forget a first vision of Venice, of Rome, of Istanbul, of Marrakesh, of Carthage, of Tangiers, of Paris, of Rio de Janeiro, of Moscow, of Sydney, of Cape Town?

There are some points I would want to add to alter, slightly, what Margaret Drabble says here. First, although cities can be experienced as large works of art, so that questions of aesthetics should be central to their planning, cities are not just works of art; whether or not we experience them as pleasant or repugnant depends on more than this. Good town planning therefore depends on more than good urban design. Secondly, and most important, Margaret Drabble speaks of *architecture* and of *buildings*. But although individual buildings, and thus architecture, are important to the quality of towns, it is the whole ensemble of buildings and spaces in a town – including its parks and gardens – which governs how we experience it (notice how in the above quotation Drabble slides from talking about individual buildings to cities as a whole). In other words (and in so far as it is an 'art'), it is really the art of town planning which emerges from Drabble's reflections as arguably the most significant art. But with these qualifications added, what Drabble says here is very important, for it draws attention to the extraordinary fact that town planning (and architecture) is not generally perceived as very significant in our society (notice, again, her initial surprise at her friend's 'eccentric' suggestion). And this even though most of us live in cities, and even though most of these cities are unpleasant to be in, and even though it is possible for humans to create wonderful cities for people to inhabit.

If town planning is as important as I contend, then clearly so too is the general theory which underpins it. Moreover, from the above discussion it would appear that a central part of that general theory should be concerned with three questions: First, what are the components of good-quality urban environments? Secondly, under what conditions are these qualities most likely to be realised? And third, to the extent that public sector town planning is one of these conditions, what part can town planning play in bringing about better cities (and a better environment more generally) for people to live in? I shall return to these questions in concluding this book. However, for much of the time since the Second World War planning theorists have been more preoccupied with other questions, and particularly with the basic conceptual question of how we should conceive of (and so *define*) the discipline of town planning. There have been good reasons for this, the chief amongst them being that theorists of town planning since 1945 have held different, and in some ways opposing, conceptions of town planning, and thus different and opposing views about the theory which is most relevant to inform it. But this is to anticipate the story which I should now begin.

NOTES

1. Other readers in planning theory include: Burchell and Sternlieb (1978), Healey, McDougall and Thomas (1982a), Paris (1982), Mandelbaum, Mazza and Burchell (1996).
2. If this account of 'planning theory' is correct, then the 'discipline' of planning theory is rather like philosophy, for philosophy asks basic, fundamental questions about the world and our place in it. Perhaps, then, the 'planning theorist' should be someone with a philosophical predisposition, and not only in the sense of asking fundamental questions about planning but also in the sense of employing the analytical rigour that is typical of the best philosophy in examining those questions.

PART I
EARLY POST-WAR PLANNING THEORY

1

Town planning as physical planning and design

INTRODUCTION

In this and the following chapter we shall be examining the view, or theory, of town and country planning which prevailed in Britain for about twenty years following the Second World War. There are two aspects of post-war planning theory which I shall distinguish and examine separately in this and the next chapter.

First, in this chapter, I examine the prevailing conception of the *nature* of town (and country) planning as a discipline; that is, the view which most town planners held in the post-war years about the kind of activity they were engaged in – how planning theorists at this time would have *defined* town planning. A useful way of approaching this is to imagine a leading town planner of the post-war years being asked by an intelligent layperson: 'what *is* town and country planning?' Although as we shall see, the concept or definition of town planning which prevailed at this time could be summarised in one or two sentences, we get a richer picture if we fill out this definition somewhat, and that, too, I shall do in this chapter.

Second, in the next chapter, I examine the main views held during the post-war period of what the *purposes* or *aims* of town planning should be. This necessarily involves an inquiry into the values which underpinned town planning at this time, and so in describing this second aspect of post-war planning theory we examine the *normative* theory of planning which predominated in those years.

First, the prevailing view held in the post-war years of the *nature* of town planning. The concept of town planning which predominated was similar to that which was held during the war and pre-war years and, indeed, long before that. During and after the Second World War there was in Britain (as in other western democracies) an added political ingredient to town planning because of the widespread discussion about establishing a new *system* of planning for the country as a whole. This was connected with a view that emerged following the war and the interwar economic depression that the state should play a much more active, interventionist role in society. The post-war Labour Government represented this emergent position of 'social democracy' (as it came to be called), and between 1945 and 1951 this government established a new political agenda based on an expansion of the state's responsibilities: a 'welfare' state providing universal education, health care and social security, etc., and in the state's more active role in managing the economy (including, in some cases, the nationalisation of major industries and services). The expansion of the state's role in town planning, as represented by various pieces of planning legislation (of which the centrepiece was the Town and Country Planning Act 1947), was thus part and parcel of this new post-war politics.

But if people had been asked at this time what sort of an *activity* town and country planning was, then, I suggest, their answers would have reflected a concept of town planning that had not changed significantly for some hundreds of years, since at least the time of the Renaissance and subsequent European Enlightenment. It was generally assumed that town planning was essentially an exercise in the *physical planning and design* of human settlements. As such, it was seen as a natural extension of architecture and (to a lesser extent) civil engineering, and hence as an activity most appropriately carried out by architects (and civil engineers). It is therefore this 'physicalist', design-based view of town and country planning which I describe in this chapter.

Before doing so, there are two preliminary points to note here which anticipate material presented later in the book. First, whilst conceptions about the *nature* of planning during the post-war years exhibited continuity with earlier periods of history, views about the *purposes* or *aims* planning should pursue were more particular to that time and had their roots in more recent history (see Chapter 2).

Secondly, though the view about the nature of town and country planning stretched back into history, it was a view that came to be questioned and to some extent abandoned during the 1960s because many of the outcomes (or apparent outcomes) of post-war planning *practice* were criticised in the late 1950s and 1960s. The conception of town planning described here is one which persisted for about twenty years following the Second World War. After that, new ideas and perspectives emerged, and it is the task of the rest of this book to describe these.

My account of the 'physicalist' conception of planning is drawn chiefly from books and other written sources published in and around the period of the Second World War, and especially from 'textbooks' which sought to explain, in a general sense, what town and country planning was about. After all, our understanding of the view of planning that was taken during this or any other period must rest to a large extent on what relevant people *said* about it, and this translates, for the most part, into what people *wrote* about planning. Examples of such texts include Patrick Abercrombie's *Town and Country Planning* (first published in 1933), Thomas Sharp's *Town Planning* (1940), Lewis Keeble's *Principles and Practice of Town and Country Planning* (1952), and Frederick Gibberd's *Town Design* (1953). Keeble's was a standard and highly recommended textbook for students and practitioners of planning from the time of its publication through to the mid-1960s, and thus it expresses in a particularly vivid way the view of town and country planning which prevailed during this period. As was written on the sleeve of the fourth edition of Keeble's book published in 1969:

> 'Principles and practice' has always been much more than a student's textbook. In this edition it emerges fully as probably the clearest and most explicit, certainly the fullest and most comprehensive, work yet published upon the vital subject of physical planning . . . Today there are few planning offices and almost certainly no schools of planning in the English speaking world where it is not in use.

The blurbs on book jackets, of course, always makes grand claims like this. Nevertheless, I do not think this particular claim is either untrue or unreasonably immodest. Throughout the 1950s and into the mid-1960s, Keeble's book was recommended to all students of planning (and often as the main course text). It was also used as a standard work of reference, even as a planning 'manual', in many planning offices, so that amongst planners themselves it was probably the best known and most widely used book on town planning.

THE COMPONENTS OF THE POST-WAR CONCEPTION OF PLANNING

The description of town and country planning in the post-war period (and long before that was conceived) as essentially an exercise in physical planning and design, but this abbreviation needs to be more fully explained. We can distinguish three related components to this:

1) Town planning as *physical* planning.
2) *Design* as central to town planning.
3) The assumption that town planning necessarily involved the production of *'master' plans* or *'blueprint'* plans showing the same degree of precision in the spatial configuration of land uses and urban form as the 'end-state' blueprint plans produced by architects or engineers when designing buildings and other human-made structures.

Town planning as physical planning

After the Second World War, there was much talk of 'planning' in a *general* sense – that is, state intervention in, and playing a more active role in, the managing and planning of social and economic affairs generally as part of the changed political climate. As town and country planning was only one form of planning activity, the question naturally arises as to what made town and country planning different from other forms of planning. The prevailing view was that, with the possible exception of regional planning controls over industry,[1] town and country planning was concerned with the 'physical' environment and was thus most appropriately described as *physical* planning, as opposed to 'social' and 'economic' planning. As Keeble (1952, p. 1, emphasis added) put it on the first page of his book:

> Town and Country Planning might be described as the art and science of ordering the use of land and the character and siting of buildings and communicative routes . . . Planning, in the sense with which we are concerned with it, deals primarily with land, and *is not economic, social or political planning,* though it may greatly assist in the realisation of the aims of these other kinds of planning.

There are three points about this statement worthy of comment. The first concerns the conceptual problem of distinguishing between 'physical' and 'social' (as well as 'economic') planning. The second concerns the alleged relation between physical and other forms of planning. And the third concerns the suggestion that town and country planning is not 'political'.

The conceptual problem arises because it is difficult to make much sense of the idea that town and country planning is *not* concerned with 'social' and 'economic' matters. One could suggest that town and country planning is concerned with the 'physical environment' – and so with buildings, roads, land, etc. (i.e. with physical *objects*), and that this is distinct from planning (for example) health care or education. The former could be described as 'physical' and the latter as 'social' planning. This is, however, a rather contrived distinction. If one were to ask what physical planning is *for*, or *why* one might wish to plan a part of the physical environment, then it is difficult to think of a reason for this planning which is not 'social': for people generally wish to control the form of their environment to maintain or enhance their well-being or welfare. The nineteenth-century town planning movement in Britain was very much concerned with the physical planning of cities for reasons of public health, and policy for health is generally regarded as 'social'. Furthermore, town and country planning is a form of social action just as much as planning the provision of health care or education. So there is some incoherence in this distinction between planning which is said to be only, or even primarily, 'physical', and planning which is, by contrast, 'social'.[2] However, as is evident from Keeble's way of defining planning, town and country planning was typically thought of at this time as being about the physical environment, and hence as only *physical* planning.[3]

This is not merely a pedantic point. For if we allow that there is some distinction between 'physical' and 'social' planning, the question of whether town planning should be defined as 'physical' (and *not* 'social'), or alternatively as 'physical *and* social', is a question of what the proper *scope*, and hence the *purposes*, of town planning should be; it is a question of whether town planning should be conceived as an activity which is 'only about' the physical environment and physical development or as a wider activity encompassing 'social' and 'economic' matters as well. Donald Foley drew attention to these alternative conceptions of town planning in a well-known paper about the ideology of British post-war planning (Foley, 1960). Here he made clear that there is considerable tension, and ideological debate between, a 'physicalist' view and a wider 'social' concept of town planning.

Secondly, Keeble suggests that town planning, though it is not social and economic (or even political) planning, 'may greatly assist in the realisation of the aims of these other kinds of planning'. If we allow that there is some distinction between physical, social and economic ends, then implicit in this statement is an assumption that social and economic ends could be advanced by physical means – that is, by the location, siting, disposition and physical layout of buildings and roads, etc. At one level there is nothing exceptional about this, for clearly the physical form and layout of a town can affect social and economic life (e.g. new roads can attract commercial development to an area; and a toddlers' play area can attract young children and so bring children in a neighbourhood into contact with each other). Keeble's statement, however, is worth attending to because the idea that the physical form of the environment could affect social and economic life was quite central to planning thought at the time. This sometimes took the stronger thesis that the physical form and layout of buildings and spaces could *determine* the quality of social or economic life, and this thesis was appropriately termed physical, architectural or environmental determinism (see Broady, 1968, Chap. 1). The post-war 'Mark 1' new towns, for example, were designed from a common assumption that, by laying out residential areas in physically distinct neighbourhoods, with 'their own' local shops, recreational open spaces, primary schools, etc., there was a greater likelihood that a 'social' neighbourhood (i.e a 'community') would develop. As it turned out, this was sociologically naive (as we shall see in Chapter 3). Nevertheless, this assumption was built into early post-war planning thought, and Keeble's statement hints at this.

The third point concerns Keeble's assertion that town and country planning is not 'political' planning. Again, much hangs on how we interpret this. If he meant that town and country planning is not concerned with planning the political *system*, then we could concur with this. But if he meant that planning does not involve or assume a commitment to a political *position*, then this is questionable. The very introduction of land-use planning entails an acceptance of some form of state intervention in the property market, which in turn entails a particular political ideology (such as social democracy). Indeed, the introduction of publicly accountable town planning presupposes, even if it does not itself directly 'plan', a certain kind of political system, so that from this point of view town planning *is* a form of 'political planning'. Decisions about how land should be used and developed necessarily involve making choices which affect the interests of different groups in different ways, and so these choices are also 'political' in this sense. Whatever Keeble himself may have meant, his statement is worth attending to because it was also part of the prevailing conception of town and country planning that planning was primarily a 'technical' activity, and so an activity that was not in itself political, or which at least did not carry with it any specific political values or commitments. Indeed, its designation as 'physical' (not 'social' or 'economic') was precisely one of the reasons why people at this time thought of town planning as technical and apolitical.

Assuming that town and country planning was conceived of as physical planning, the question naturally arises as to what technical skills were thought relevant, which brings us to the second component of the post-war conception of planning.

Town planning as urban design

Because town planning was viewed as an exercise in planning the physical location, form and layout of land uses and buildings, it was also regarded as an exercise in physical or urban *design* (the term 'civic' design was also much used). Town planning was regarded as an 'extension' of architectural design (or to a lesser extent civil engineering) in the literal sense of being concerned with the design of whole groups of buildings and spaces – with 'townscape' rather than the design of individual buildings and their immediate sites, and also in the sense that architecture too was seen to be an exercise in the *physical* design of built forms. It followed that the professionals generally considered as most qualified to undertake such work were architects, together with the two other main built environment professions, civil engineering and surveying. It is not surprising, therefore, that these other built environment practitioners resisted the establishment of town planning as a separate profession in Britain on the grounds that town planning was a natural extension of their work and hence part and parcel of their brief (see Cherry, 1974, Chaps. 6 and 7). Hence, although the British Town Planning Institute had been established in 1913 and had petitioned for a Royal Charter in 1948, it was not until 1971 that the institute succeeded in obtaining the grant of a Royal Charter to become fully recognised as a distinct profession.[4]

Most practising town planners in the immediate post-war period, therefore, were architect-planners.[5] Three of the most famous planners in post-war Britain – Patrick Abercrombie, Frederick Gibberd and Thomas Sharp – were all architects. This situation was reflected in other European countries: in The Netherlands, for example, from the end of the First World War to the mid-1930s, the early modernist architect H.P. Berlage was responsible for Amsterdam's southern extension plan, and in the post-war years the famous modern architect Le Corbusier was commissioned by various cities to prepare town planning schemes.

It is thus not surprising that most of the town planning treatises written at the time put great emphasis upon urban design. Books written specifically about urban design, such as Frederick Gibberd's *Town Design* (published in 1953), were regarded as standard texts on town planning. And in Europe generally, most of the influential twentieth-century tracts on town planning, such as those by Tony Garnier (1917) and Le Corbusier (1924; 1933), likewise saw the task of planning cities as an exercise in large-scale urban design.

This emphasis on town planning as urban design is very evident in Lewis Keeble's *Principles and Practice of Town and Country Planning*, as Figures 1.1–1.5 show. 'Theoretical' master plans for new towns (Figure 1.1) are worked up by the author into a detailed design (Figure 1.2). Then, homing in on particular areas within an imaginary town, there are examples of representative designs for a town centre (Figure 1.3, also reproduced on the cover of the 1969 edition), and for residential neighbourhoods (Figure 1.4). Even a plan for an imaginary urban region is shown as if it were an exercise in large-scale design (Figure 1.5).

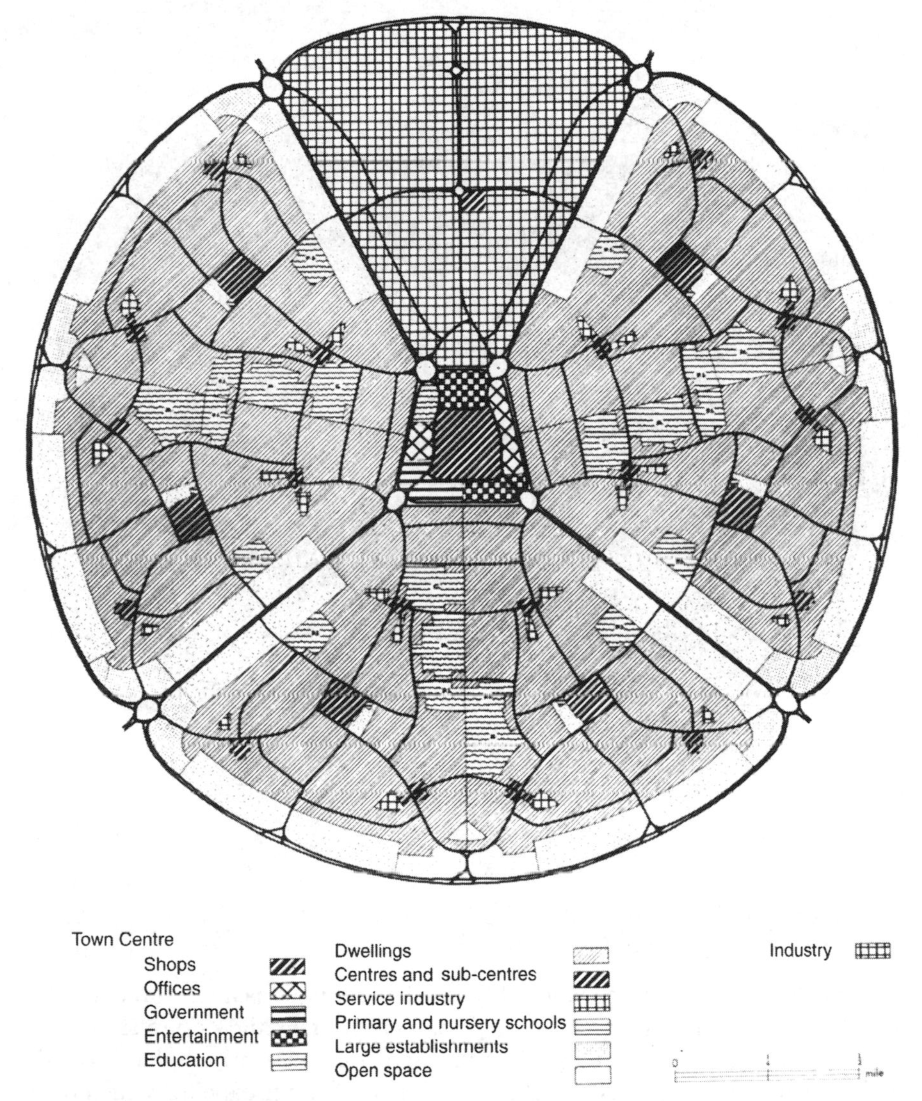

Figure 1.1 Theoretical new town
Source: Keeble, 1952 (1969 edn), Figure 30

The training town planning students received naturally reflected this approach. 'Studio work' (design projects of various kinds – housing layouts, designs for shopping centres, town centre plans, master plans for imaginary new towns, etc.) was at the heart of planning education, and all students were equipped with the same kind of drawing materials as architectural students (drawing boards, T-squares, set-squares and scales, cartographic pens and pencils, Letraset for printing, etc.). There were differences. Whereas architecture students were engaged more directly on the detailed design work for individual buildings, town planning students were concerned with the design of whole groups of buildings and urban spaces – in other words, with design

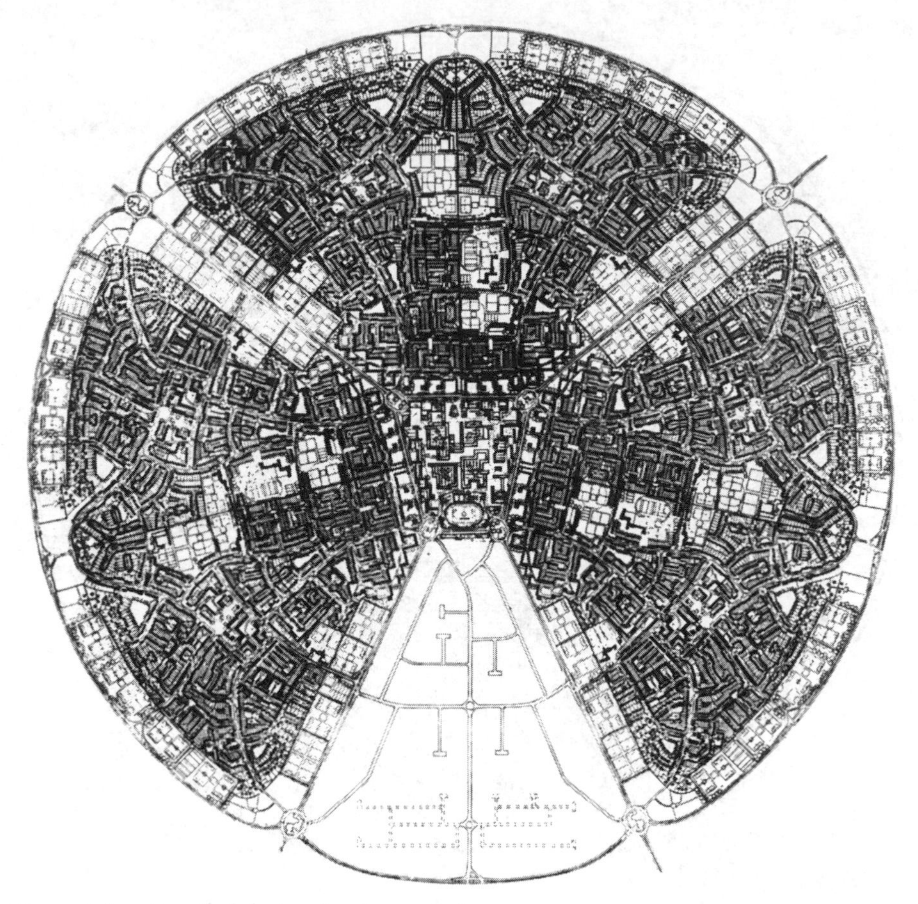

Figure 1.2 Detailed design for theoretical new town
Source: Keeble, 1952, Figure 31 (1969 edn)

'layouts'. But town planning was still viewed and taught as a natural extension of architectural training, involving the same kinds of spatial design skills.

With this emphasis on town planning as design was an emphasis on the *aesthetic* character and qualities of existing areas of townscape for which plans might be prepared, together with an emphasis on making plans which (it was hoped) would enhance the aesthetic quality of environments. Raymond Unwin – a leading exponent of this concern with aesthetics – stressed the need for beauty in urban life: 'Not even the poor can live by bread alone' (cited in Creese, 1967, p. 71). Unwin spoke of town planning unreservedly as an 'art' which would provide 'the opportunity of a beautiful environment out of which a good human life would grow' (Unwin, 1930, cited in Creese, 1967, p. 165; note again the physical determinism of this). The aesthetics of urban form and design dominated the standard post-war texts on town planning. Thomas Sharp's (1940) *Town Planning*, for example, was greatly preoccupied with the aesthetic qualities of suburban as compared with terraced housing development.

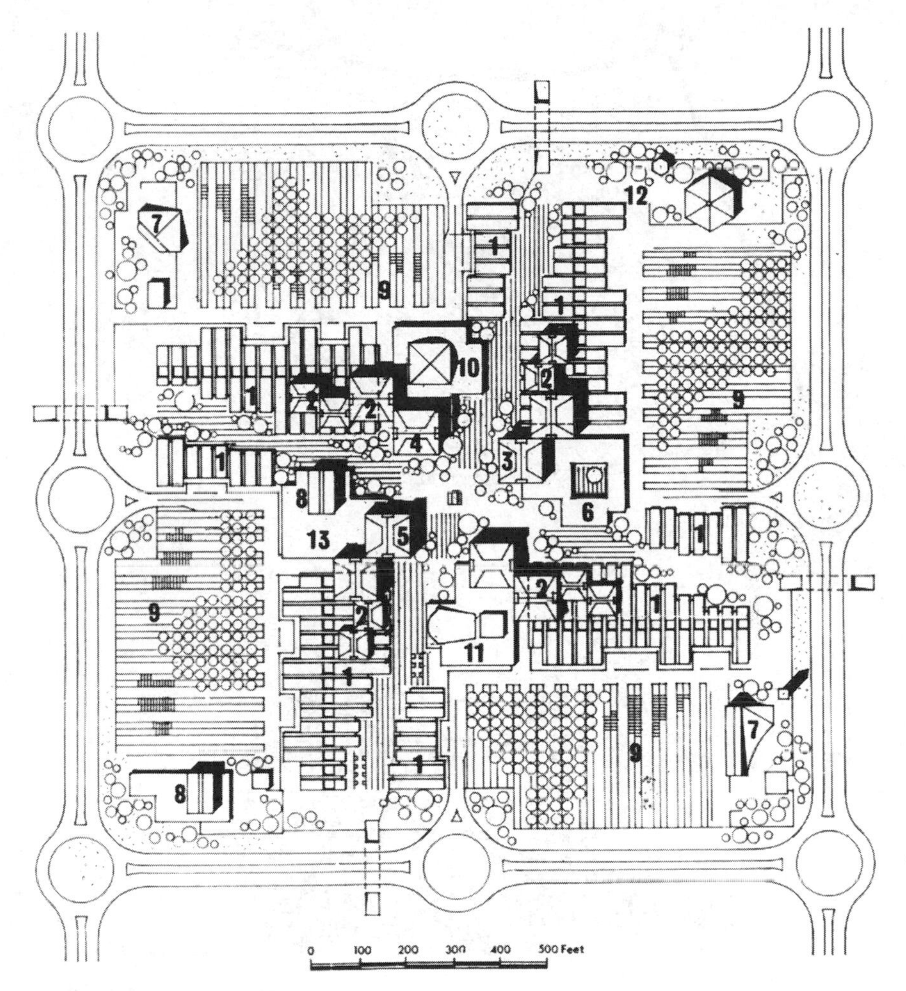

Figure 1.3 A design for the centre of a theoretical new town
Source: Keeble, 1952, Figure 78

The centrality of aesthetics is also echoed in many of the town planning reports produced at the time, and the planning reports for cities produced by Thomas Sharp provide a vivid illustration of this. Sharp's (1946) plan for the blitzed city of Exeter begins with an analysis of the aesthetic character of the city centre and his proposals are largely governed by aesthetics. For example, his block design for the pedestrianised shopping street Princesshay was located and aligned on aesthetic grounds to provide a perspective view of the cathedral, rather than from an analysis of the locational requirements of retail businesses or of people's shopping behaviour.

Admittedly, land and buildings were used and thus how the parts of a town 'functioned' were also considered as part of the process of urban design. This was, after all, the age of modernist 'functional' architecture. Just as architects saw architectural design as the art of designing forms to accommodate (even

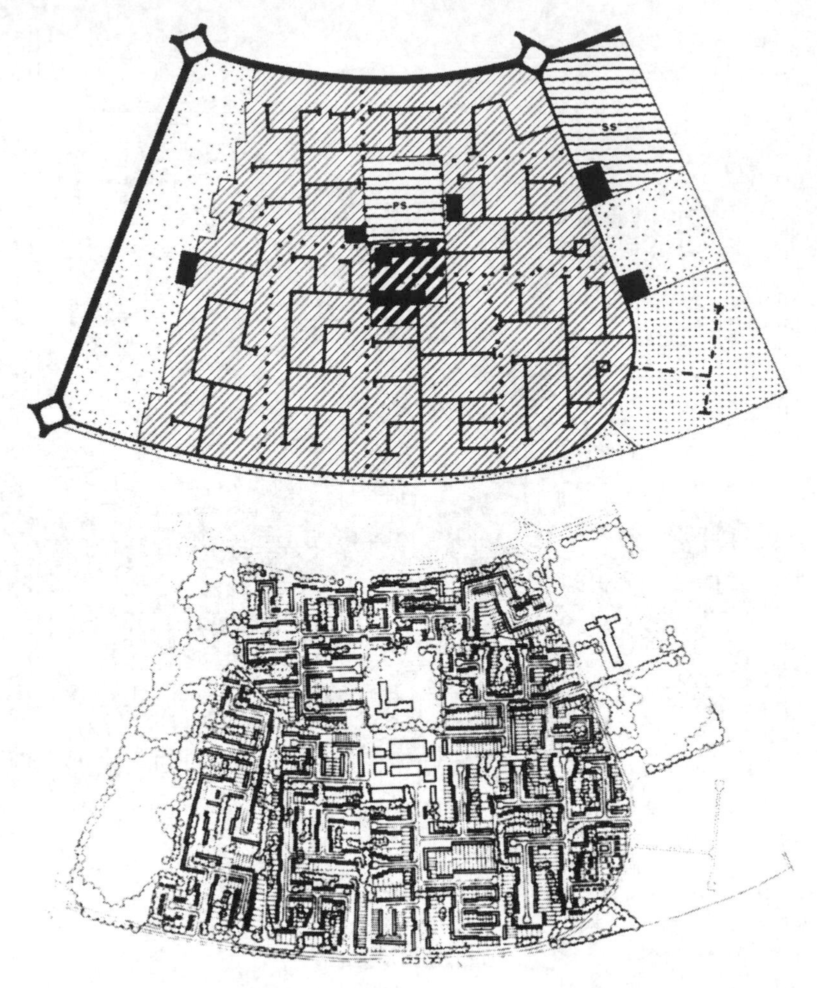

Figure 1.4 Design of neighbourhoods in a theoretical new town
Source: Keeble, 1952, Figures 93 and 94, 1969 edn.

to 'follow') functions, so urban designers (most of whom were architects anyway) saw their job as one of designing functionally efficient towns. In spite of this, however, a kind of aesthetic formalism dominated ideas of the well designed town so that urban functions were visualised and accommodated for in terms of some prior aesthetic conception of the ideal urban form rather than the other way round (this was also true of much modern architecture, notwithstanding the rhetoric of functionalism). For example, in deciding where industry should be located, a prime consideration was to site it away from residential areas. It was thought (no doubt correctly in many cases) that it was unpleasant to live next to factories, but this notion tended to prevail to the locational requirements of the industries themselves which, in any case, were hardly taken into account by planners, bound as they were to a design conception of planning. This approach is all too evident in Keeble's plan for a

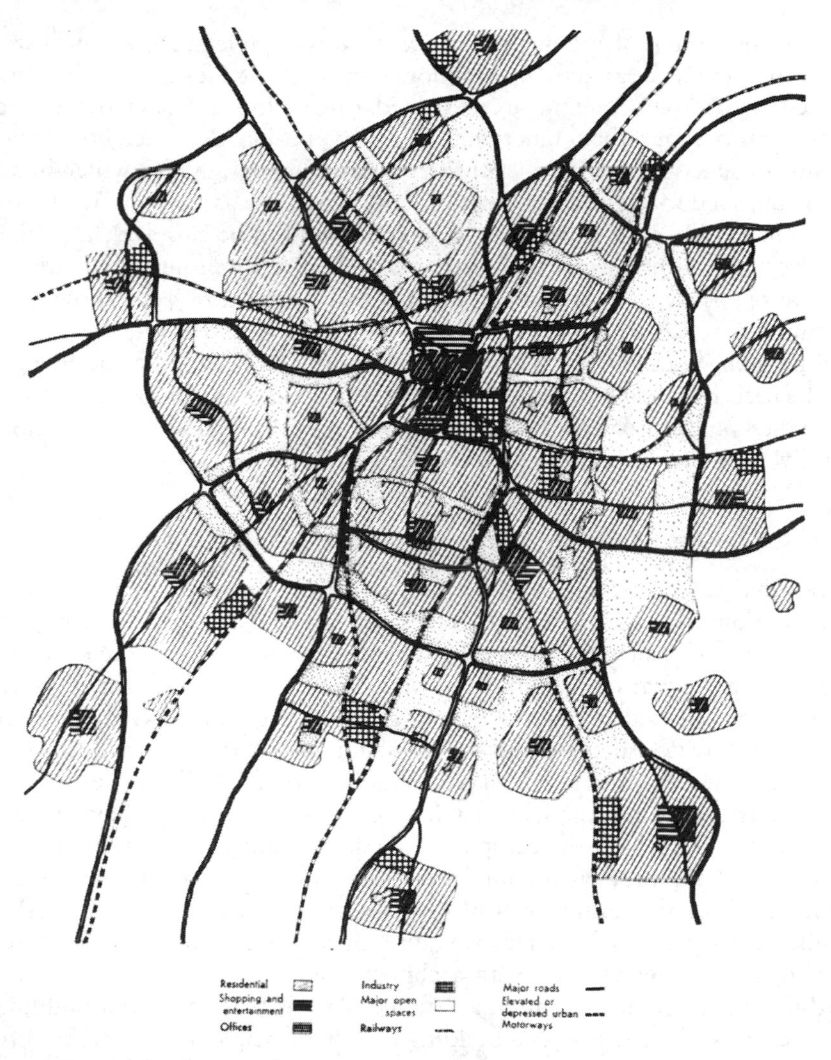

Figure 1.5 A plan for an urban region
Source: Keeble, 1952, Figure 11

theoretical new town (Figure 30 in Keeble, 1952), where all industry is collected into one sector of the town. This model was adopted for most of the 'Mark 1' new towns built in Britain after the war (though typically with two industrial zones rather than just one) and, from a traffic point of view, this proved to be anything but functionally efficient: each morning and evening peak traffic tidal flows emerge from one or two sectors of the town.[6]

In the residential planning of new towns, plans were laid down on the basis of a neat pattern of physically distinct neighbourhoods, all of which were roughly the same size, with same-sized local shopping centres, primary schools and the allocation of green open space. As such writers as Christopher Alexander (1965) and Maurice Brown (1966) were later to point out, this 'tidy design' conception of urban form showed no real understanding of how different residential areas actually functioned or, in what Alexander called 'natural' cities, of how different areas tended to develop differing patterns and concentrations of urban functions. The very notion that 'neighbourhoods' existed or functioned as distinct entities was itself a design idea which had not been subjected to critical examination and, when it was (e.g.: by Ruth Glass, 1948, in a study of Middlesbrough), it was found to be 'suspect'. During this postwar period there was hardly any systematic research into how cities and regions or specific activities functioned. As Eric Reade (1987) has shown, town planning practice was not grounded in empirical research and theory. Plans and planning decisions were made largely on the basis of intuition or, rather, on the basis of simplistic aesthetic conceptions of urban form and layout which embodied physical determinist assumptions about how best to accommodate the diverse economic and social life of cities.

Town plans as detailed blueprints or 'master' plans

Given that town planning was viewed as essentially an exercise in physical design, along the lines of the architectural model of design, it seemed self-evident to town planners at this time that their prime task was the production of *plans* – town plans, regional plans, plans for village extensions and so on. It also seemed self-evident that these plans should be as detailed as far as possible to guide future development and should define, as precisely as feasible, sites for particular uses. In other words, a plan should in principle show the extent and form of that town at some specified date in the future when, all being well, the plan would be realised or 'completed' (little thought was given at this time, however, to the problem of implementing such plans). Plans were seen as 'blueprints' for the future form of towns – as statements of 'end-states' that would one day be reached. This was directly analogous to the work of architects or civil engineers, where an architect's or engineer's design would also lead ultimately to the making of a detailed plan or blueprint for a building or some other structure. Just as a building can, in principle, be constructed from an architect's final drawings, so too could a town be developed by reference to its master plan, at least in terms of the 'outlines' of development.[7] This is hardly surprising, given that most of the town planners were architects by training. As Peter Hall (in Hall *et al.*, 1973, Vol. 2, p. 38) was later to write: '. . . the planning profession in Britain from the outset had a heavy design bias: an obsession with plan design, in the form of a physical blueprint.'

The 'blueprint' character of town plans is well illustrated by the plans for new towns which were produced in the 1940s and 1950s. Indeed, the end state character of these plans was evident in that it was generally assumed that a time would come when the new town's development, based on its plan, would be completed. The first generation of development plans local authorities were required to produce under the Town and Country Planning Act 1947 also adopted this approach. Detailed zoning plans specified how particular sites were to be used and developed, including the detailed alignments of any road

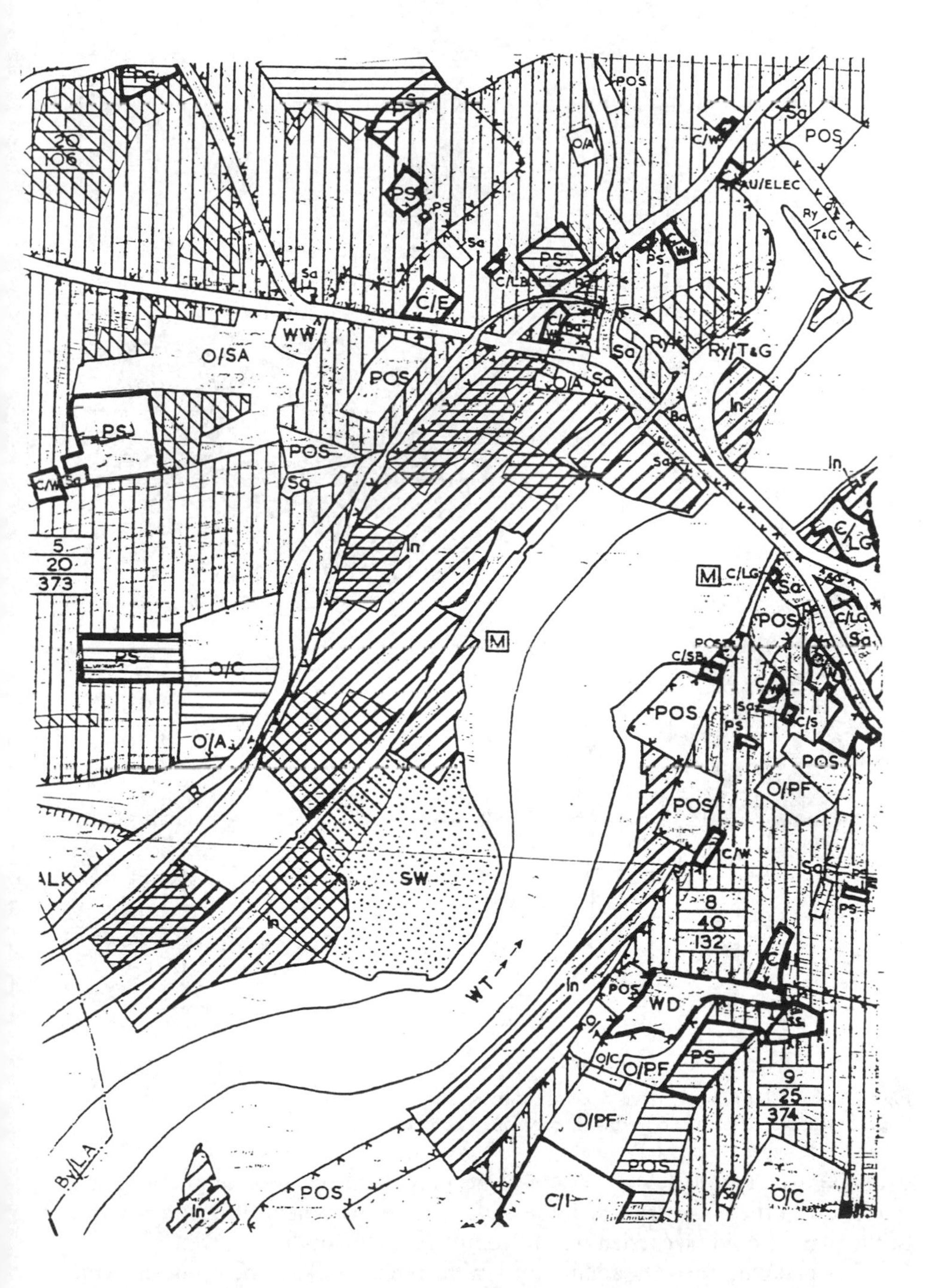

Figure 1.6 Part of a development plan produced under the Town and Country Planning Act 1947

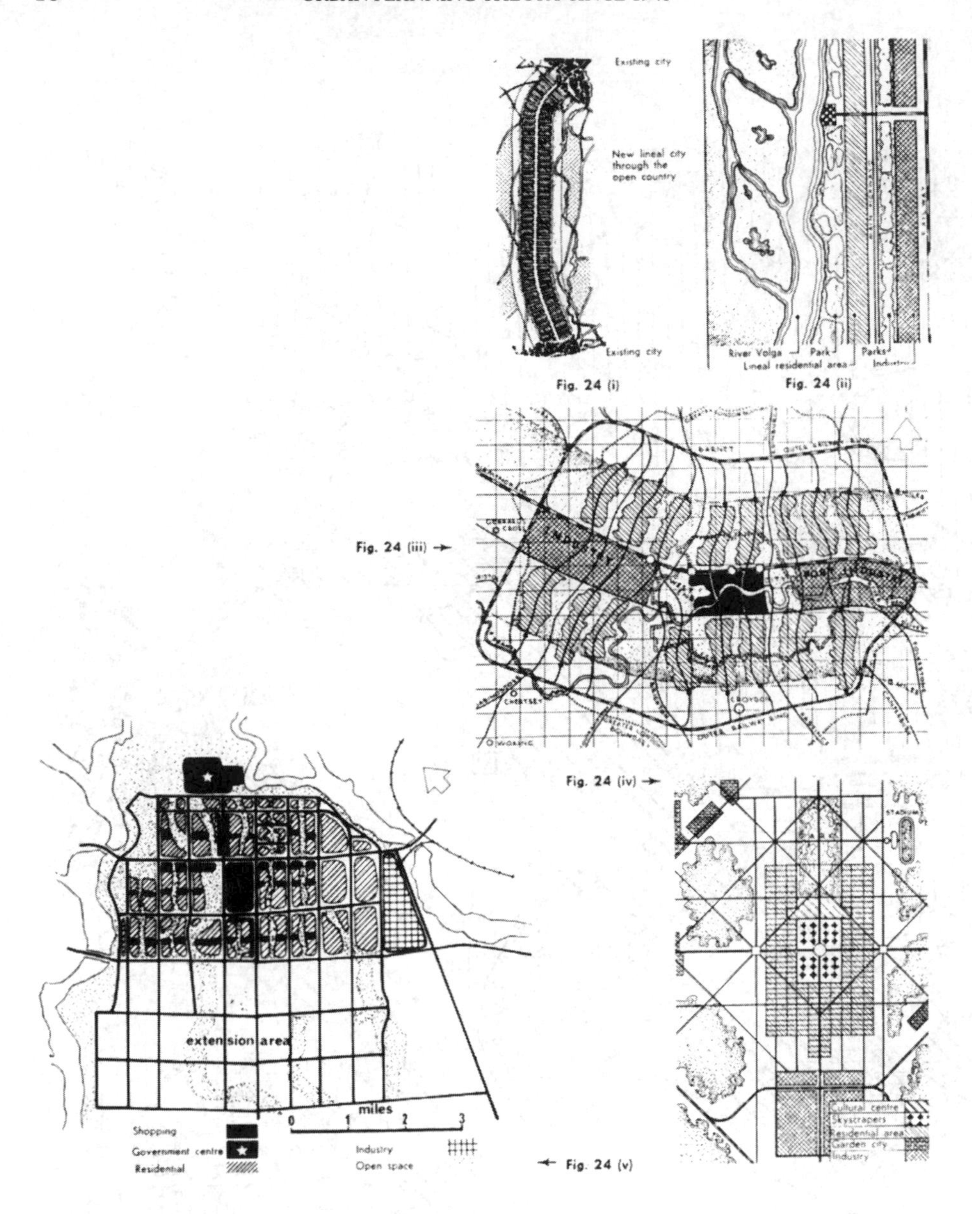

Figure 1.7 Planning theories according to Keeble
Source: Keeble, 1952 (1969 edn), Figure 24, 1969 edn.

widening (see Figure 1.6). These were accompanied by 'programming' plans that showed the stages at which the envisaged development of different parts of the plans would be carried out to 'complete' the plans.

Town planning theories at this time were often similarly preoccupied with visionary plans or designs that showed how the ideal town or city should be spatially organised. Soria y Mata's nineteenth-century plans for linear cities, Le

Corbusier's plans for the 'contemporary city' (and later the 'radiant city') in the 1920s and 1930s, and Frank Lloyd Wright's plans for 'Broadacre City' in the 1930s were exemplars of this approach, as had been Ebenezer Howard's 'Garden City' (with some important qualifications about Howard's land policy proposals). In other words, because town planning was viewed at this time as an exercise in planning and designing the physical form of towns, it was natural that theories of planning were expressed as master plans for urban form. The point is well illustrated by Figure 1.7, from Keeble's standard text (Keeble, 1952). The figure shows five different suggestions for how the form of towns might be planned. Each is essentially a master plan, or a blueprint, and it is significant that Keeble labelled this figure 'Planning theories'. For the idea that town planning was essentially about physical design, and hence involved producing blueprint plans for future urban form, was precisely the conception, or theory, of town planning which prevailed at the time when Keeble wrote his book.

CONCLUSION

In this chapter I have described the main features of the conception of town and country planning which held sway during the years following the Second World War in Britain. In fact, this conception – which viewed town planning primarily as an exercise in the physical planning and design of land use and built form – long pre-dated the Second World War, even the twentieth century. Thus for as far back as the European Renaissance, town planning – in so far as it had been conceived as a separate enterprise at all – had been seen as a natural extension of architecture onto the larger stage of urban streets and piazzas. But really, until the late nineteenth century, town planning had *not* been distinguished from architecture, and precisely because it was seen as architectural design on a larger canvas.

The view of pre-war planning described in this chapter was not peculiar to Britain. Since the town planning movement in this period was very much an international movement, the idea that town planning was 'like' architecture, but on a larger scale, was also the norm in other European countries and in North America. The European concept of town planning, which has design at its heart, has proved more durable than that in Britain, as has the idea that the product of town planning should be detailed land-use zoning or 'master' plans. In France the local 'POS' plans ('plans d'occupation des sols') are still essentially detailed 'blueprint' plans, as are the local 'binding' plans (the 'bestemmingsplans') in The Netherlands. Furthermore, in Britain, the idea that urban design is central to (even if no longer definitive of) town planning experienced something of a resurgence in the late 1980s (see Chapter 8). The concept of town planning as physical design has not, therefore, been entirely discredited or superseded.

However, one thing which has changed since the early post-war years has been the kind of visionary Utopianism that characterised the expression and promulgation of many town planning ideas throughout the first half of the twentieth century (Fishman, 1977). Implicit in these imaginary master plans –

whether they were the grandiose plans of Le Corbusier for the 'radiant city', or the more prosaic plans we find in Keeble's textbook – was not just an approach to town planning as an exercise in physical planning and urban design but also a normative concept of the ideal urban environment. In other words, the tracts and textbooks published at the time not only advanced an extended *definition* of planning but they also embodied certain *values* about the kinds of environment which, it was believed, should be realised through town planning. It is to the normative ideals of post-war planning theory that the next chapter turns.

NOTES

1. Before, during and after the Second World War, town and country planning was seen by some to include regional planning as well as the more specific task of planning towns. So even at this early date there was a concern with *strategic* as well as local land-use and development planning. This wider perspective can be traced back to Geddes's writings (e.g. Geddes, 1915). It was exemplified by the report of the Barlow Commission (1940) which was concerned centrally with the regional pattern of economic activity and employment, as well as the physical extent and pattern of urban development. The Barlow report was the basis for the first of the major post-war planning Acts, the Distribution of Industry Act 1945, which initiated controls over the regional location of industry. Strategic-level planning was, then, seen to be an aspect of 'town and country planning'. However, this level of planning was clearly distinguished from the localised form of town and country planning established by the Town and Country Planning Act 1947. Thus post-war regional planning controls over industry were administered by the Board of Trade, not the ministry responsible for town and country planning.
2. The same problem arises with the distinction between 'physical' and 'economic' planning, and indeed between 'social' and 'economic' planning.
3. The distinction persists to this day in the difference often made in official circles between so-called 'planning' or 'land-use' matters, and 'social' or 'economic' matters.
4. The Town Planning Institute's petition for a Royal Charter in 1948 was met with counterpetitions from the Royal Institute of British Architects (RIBA), the Royal Institute of Chartered Surveyors (RICS) and the Institute of Civil Engineers (ICE). Each opposed the TPI's petition on the grounds that town planning was not separate from their own areas of work. As the ICE petition put it: 'Town planning is not a separate professional activity. It requires the bringing together of the art of the architect with the skill of the civil engineer and the knowledge of the surveyor, as well as a knowledge of public administration, a judgement of economic trends and population distributions and other matters of a social character' (cited in Cherry, 1974, p. 167). Note, too, how this statement emphasises physical design skills, and how the skills of incorporating 'social and economic' issues into town planning are very much tacked on.
5. To be precise, over the period of 1946–56, 45% of the Associate Members of the Town Planning Institute were architects, as compared with 22% 'direct entry' planners, 14% engineers, 9% surveyors, and 9% holding other first-degree qualifications, such as geography (figures from Cherry, 1974, p. 210).
6. This was one of the things which distinguished the post-war 'Mark 1' new towns from later new towns, such as 1960s Milton Keynes, throughout the town to spread journey-to-work traffic more evenly over the road network.

7. Decisions about the detailed architectural form and style of new development were of course beyond the scope of a land-use zoning plan alone. This is why the first generation of these plans produced following the Town and Country Planning Act 1947 were very poor instruments for achieving good-quality design, even though good design was a central concern of town planning.

2

The values of post-war planning theory

THE NORMATIVE CONTEXT: A CULTURE OF SOCIAL REFORM AND CONSERVATIVE SENTIMENTS

Town planning as an exercise in physical planning and design represented a particular theory of what kind of an activity town planning is – in other words, a 'formal' or 'definitional' theory of planning. However, British post-war planning was also driven by a distinct set of *values* which, when generalised, amounted to a *normative* theory of what constituted the ideal physical environment which it should be the task of town planning to try to bring about. The internationalism of town planning as physical planning and design was emphasised in Chapter 1. However, the *values* which drove British post-war planning were more particular to that country, to its time and place. They reflected the responses of social reformers and middle-class intellectuals to the dreary industrial cities which had grown up in the Victorian age, and were a curious mixture of radicalism and conservativism.

The grim living conditions endured by working-class people in British nineteenth-century industrial cities had generated a movement for radical social reform, and this dovetailed with the rise of socialism. Thus Robert Owen – the creator of the famous model settlement of New Lanark – was both a pioneer of the model village movement, which aimed to improve the living and working conditions of working-class people, and an early socialist. The model village and socialist movements were characterised by a radical Utopianism which sought a completely new kind of urban settlement to accommodate a new kind of society. Ebenezer Howard's ideas for the creation of completely new 'garden cities', in which land would be collectively owned, came to represent at the end of the century the distillation and most complete expression of this radical Utopian socialism.

British Utopian socialism, on the other hand, was also characterised by a romantic vision of a pastoral, preindustrial past, and this nostalgia for the past had affinities with conservative sentiments. This is evident in the socialist writings of William Morris, whose ideas about art and craftsmanship and whose vision of a future urban Utopia described in *News from Nowhere* looked back longingly, and romantically, to medieval England (Morris, 1890). This backward-looking romanticism is also evident in Howard's *Garden Cities of Tomorrow* (Howard, 1898). Howard combined radical socialist proposals for the collective ownership of land in his garden cities with very traditional and, in this respect conservative, notions of urban size and form. The designs of Howard's 'garden cities', with their curvaceous tree-lined avenues of neo-gothic cottages, represented an expression in town planning of that more general neo-gothic romanticism that infected the whole of Victorian Britain. The garden city planning movement – which was to have such a profound influence on British town planning in the twentieth century – thus illustrates perfectly this mixture of radical and conservative values.

The heir to this tradition of radical reform combined with conservative sentiments was the consensus politics that succeeded the Second World War – the so-called 'social democratic' consensus which was to endure for thirty years until the mid-1970s. At one level, the post-war Labour Government represented a radical politics expressed through increasing the role of the state in society in both the 'social' sphere (the so-called 'welfare state') and also in the sphere of managing the economy (including nationalisation). The town and country planning legislation introduced by this government – the New Towns Act 1946, the Town and Country Planning Act 1947 and the National Parks and Access to the Countryside Act 1949 – was consonant with this broader programme of increased state intervention.

Radical though this programme was at the time, in the pure sense of the term this programme was not 'socialist'. In its acceptance of a 'mixed' economy, in which an enlarged public sector co-existed with free market capitalism – in which, in effect, the state 'managed' capitalism but where private enterprise remained the central driving force in the economy (including land development) – the post-war government's stance is more correctly characterised as 'social democratic' rather than 'socialist'. This social democratic style of politics also commanded the support of the conservatives as well as 'radicals' following the experience of economic depression and unemployment in the 1930s and the acceptance of a Keynsian economic strategy to address these problems, and following also the necessity and apparent success of a state run economy during the war itself. There was also the need for reconstruction following the war. So had a Conservative government been elected in 1945, it is likely that it would have introduced a planning system similar to that introduced by Labour (on this, see Cullingworth, 1975, Vol. 1; Ward, 1994, Chap. 4).[1]

The consensus politics of post-war Britain was, therefore, characterised by both radical and conservative ideas to construct a 'middle-way' between the extremes of liberalism (with its support of private enterprise and the free market) and socialism (with its advocacy of greater public ownership and control). This 'middle way' was perfectly illustrated by the post-war British planning system brought into being by the Town and Country Planning Act 1947. To create a 'positive' system of planning in which the state could properly plan urban land use and development, some socialists argued, the state would have to secure control over the land market by, for example, nationalising land. Indeed, the wartime Uthwatt Committee, which had been established to examine the issue of compensation and betterment in relation to land-use planning, came close to recommending this (Uthwatt Committee, 1942).[2] However, leading Labour politicians recognised that land nationalisation would be politically unpopular, and so, in the 1947 Act, they drew back from nationalising land (or the land development industry), and instead nationalised the *right* to develop land. This was a clear piece of 'mixed economy' legislation with, on the one hand, the retention of a market system based on private *ownership* rights in land and property and, on the other, the regulation of this market by the state's control of *development* rights.[3]

One other factor should be mentioned in the normative context of British post-war planning, which tilted the balance between radicalism and conservatism in favour of radicalism, and this was the widespread acceptance by the architectural establishment of the 'modern' style in architecture and urban design. This certainly did mark a break with the past. In the interwar years, modern architectural styles had been widely adopted in Europe (e.g. for social housing in Germany and The Netherlands) but at this time British architecture and urban design were still dominated by traditional 'revivalist' styles, such as the 'Tudorbethan' semi-detached housing which characterised inter-war suburban development (see Punter, 1986; 1987; Ward, 1994, Chap. 3). However, all this changed following the Second World War, when modern architecture was suddenly embraced enthusiastically by architects and planning committees. This change can be seen in many of the planning reports produced immediately after the war, such as Thomas Sharp's report on the blitzed city of Exeter (Sharp, 1946). During the war, Exeter lost some of its most lovely pieces of architecture and urban design, such as the fine Georgian Bedford Circus and part of Southernhay West. In his report, Sharp raised the question as to whether these architectural gems should be rebuilt, but he resisted this, arguing instead that the blitzed areas of the city should be rebuilt in the style of the modern age (*ibid.* pp. 87–88, 109). Sharp's view was typical of the time and reflected the optimism of post-war Britain. Following the deprivations of two world wars and the interwar depression, there was an upsurge of energy and confidence in the capacity to build a better future without the need to rely on the past. Support for more active state intervention and 'planning' generally was part of this post-war optimism; the belief that, having achieved a victory in the war, the state and proper planning could 'win victory in peace' as well. But part of this optimism was also the willingness to embrace the new architecture.

The 'normative background' of post-war Britain was, then, a mixture of both radical and more traditional, conservative values, and both these elements informed the aims of British town planning in the post-war years.

THE NORMATIVE THEORY UNDERLYING BRITISH POST-WAR PLANNING

A normative theory of town planning can comprise two things: a theory of how town planning should be approached, and a theory of the kinds of urban environments town planning should seek to create. It is normative theory of the latter kind with which this chapter is most concerned, although something about the general approach to creating better cities which was assumed by British post-war planning will also be touched upon. In explaining the normative theory of British post-war planning, the values which informed it will also be described. By 'values' here, are simply meant those things which are deeply cherished (which are, literally, highly 'valued') and which have a bearing on how we judge the quality of urban environments. The deep values we hold often take the form of taken-for-granted assumptions and norms and, because of this, our values are often not explicitly articulated or analysed. This is certainly true of the values that underpinned British post-war planning. However, these values become apparent when we examine the kind of urban environments that were judged by planners at the time to be of high quality or 'ideal'.

Four broad planning principles of the time may help to describe this. The first concerns the general approach to creating better cities that was adopted by British post-war planning, as opposed to the kind of urban environment that was thought to be ideal. This approach can be described as *'Utopian comprehensiveness'*. As noted in the introduction to this chapter, the town planning movement in the nineteenth century was strongly influenced by Utopian schemes for new settlements. This Utopianism persisted into the twentieth century, where two visions of the ideal urban future were especially influential – Howard's scheme for garden cities and Le Corbusier's 'radiant city' (see Fishman, 1977).[4] A feature of these Utopian schemes was that they envisaged the future ideal city as a *whole*. In other words, the Utopianism of post-war planning thought went hand in hand with a 'comprehensive' approach to planning cities.

The second and third normative principles of post-war planning concern, more specifically, the view taken of the ideal urban world town planning should try to bring about. The first of these concerns the general aesthetic values which informed British post-war planning (and, as we noted in Chapter 1, aesthetic considerations were central to the whole conception of town planning at this time). These aesthetic values can be described under the heading *anti-urban aestheticism*. The third principle concerns the view most town planning theorists took of the ideal urban structure, namely, a *highly ordered view of urban structure*.

The fourth concerns a general assumption that all these principles were self-evident and thus 'commonsense' principles in themselves, commanding a consensus amongst all sections of the population. In short, there was an *assumed consensus over the aims of planning*.

Utopian comprehensiveness

Utopian schemes for ideal settlements had a long tradition stretching from Thomas More's *Utopia* in the sixteenth century (More, 1516) to William Morris's *News from Nowhere* in the nineteenth (Morris, 1890). More derived the word 'Utopia' from the Latin word for 'nowhere', and Utopian thought has ever since been characterised by turning attention away from the world as it actually is to construct an imaginary ideal world which it would be desirable to bring into being. Town planning continued to be influenced by this kind of Utopianism well into the twentieth century, and two schemes in particular had an especially powerful influence on British town planning in the 1950s and 1960s: Ebenezer Howard's proposals for garden cities and Le Corbusier's imaginary sketch of the 'radiant city'.

Implicit in both these proposals was the Utopian suggestion that town planning should turn its back on existing cities and create an entirely new kind of urban settlement, although clearly there was debate as to whether this should be Howard's garden city or Corbusier's 'radiant city'. Howard's ideas, in size and urban form, underlay Abercrombie's 1944 plan for Greater London, with his proposal for a ring of relatively 'self-contained' and 'socially balanced' new towns circling London's green belt. More generally, the aesthetic of the 'garden suburb' (which also owed much to Howard via Raymond Unwin's Hampstead Garden Suburb) was the model for much post-war suburban housing, just as it had been in the interwar years. By contrast, in the post-war slum clearance and comprehensive redevelopment schemes of many inner-city areas, it was Le Corbusier's vision of the modern city of tower blocks which arose from the rubble in the late 1950s and 1960s. However, just as Howard's original ideas were corrupted, so were Le Corbusier's. Only in the London County Council's Alton Estate bordering Richmond Park was Le Corbusier's dream of the 'city in the park' realised. Elsewhere, high-rise council blocks were set coldly on a stage of grey concrete. Such has been the fate of so many twentieth-century Utopias. The fact remains, however, that these visionary urban Utopias supplied the inspiration for much post-war planning, not only in Britain but also in much of Europe and in North America.

As noted above, Abercrombie's plan proposed *new* towns around London, whilst in inner-city areas modernist high-rise estates were erected in areas subject to *comprehensive* redevelopment. This highlights a related normative assumption of much British post-war planning, which was that town planning should plan towns (or large parts of towns) *as a whole*, or *comprehensively*. This comprehensiveness was part of the Utopian tradition: Howard's and Le Corbusier's schemes were for whole cities and sets of cities. And this 'grandiose' approach to town planning comes through in Lewis Keeble's (1952) influential textbook, *Principles and Practice of Town and Country Planning*, in which many of Keeble's hypothetical plans are plans for *whole* towns or regions.

Three further aspects of the post-war 'Utopian comprehensive' approach to planning are worth commenting on here. First, with the exception of Howard's garden city, which was very English and traditional in its urban form, the urban Utopias were typical expressions of modernist 'functional' design and aesthetics (e.g. Antonio Sant'Elia's *La Citta Nuova*, Tony Garnier's *La Cité Industrielle* and Le Corbusier's *Ville Radieuse* – see Figures 2.1 and 2.2). Thus the city of the future was ordered into great blocks or zones of single land uses, with fast motorways like great arteries connecting up the different districts. In appearance, the form of the modern city was one of plain, geometrical, 'functional' buildings standing at regular intervals in a sea of 'free-flowing' space.

Figure 2.1 Antonio Sant'Elia's sketch for *La Citta Nuova*, 1914

Secondly, the urban Utopias were modernist not only in their aesthetic appearance but also in the radical sense of presuming the wholescale clearance of existing cities, or large parts of them, to make way for the new. This again displayed the 'comprehensivist' ideology of modern planning which was most vividly promulgated by Le Corbusier, who prepared plans for a host of major cities across the world in which he openly recommended sweeping away whole areas of traditional, densely packed streets and their replacement with his open-plan 'radiant city' (Le Corbusier, 1933, Chap. 6). Two of the principles enunciated by Le Corbusier were 'The plan must rule' and 'Disappearance of the street' (*ibid.*, p. 7).[5]

Planners were realistic enough to recognise that the world could not be completely reconstructed anew, and that for the most part they had to deal with cities as they found them. Nevertheless, there was a profound belief that the large industrial cities which had grown up in the nineteenth and early twentieth centuries were so deficient as settlements that, over large areas, it was better to clear everything away and plan 'from a clean slate'. As Alison Ravetz (1980, pp. 23,40) has put it:

> The result was a 'clean sweep' philosophy of planning to which all the parties that were in any way concerned with the built environment subscribed. A few were sensible enough to see that planners could never in fact start entirely from scratch, but in the 1940s future possibilities were dazzling enough to encourage the hope that somehow this might be so.

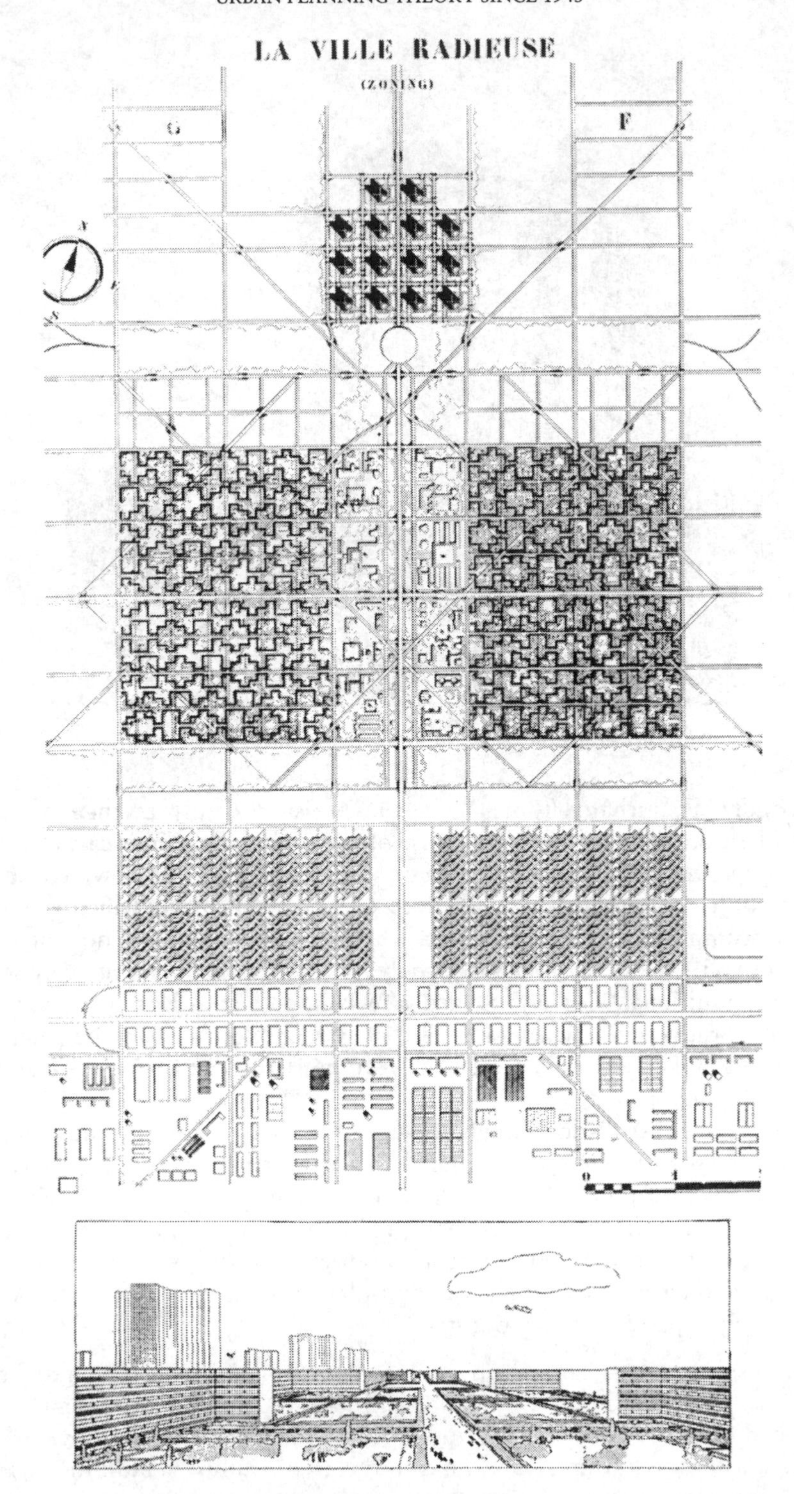

Figure 2.2 Le Corbusier's *Ville Radieuse*, plan and sketch, 1933

> For one distinguished writer on planning, it was deplorable that planners were not 'always able to detach themselves from the actual environment and to behave as though they have a clean slate and all eternity before them' . . . Central to its [i.e. post-war planning's] beliefs was the idea that the old unplanned environment could and should be replaced by a 'Utopia on the ground'.

Thirdly, because it was assumed that town plans should be comprehensive, the development plans prepared inevitably contained a whole series of proposals bound together in a single package. Town planning did not necessarily have to take this approach; it is possible, for example, to devise particular policies for particular activities (such as transport or, within this, specific policies for public transport, cycling, etc.), or policies to encourage or limit certain kinds of development in certain kinds of areas and so on. As we shall see later (in Chapter 7), the comprehensive model of planning came to be criticised. But at the time, any town plan contained a compound of many proposals. This was partly because, as we saw in Chapter 1, town plans were conceived as overall blueprints for the development of a town as a whole – as 'master' plans 'covering everything'. But it was also because it was assumed that planning *should* be comprehensive. As well as being seen as a designer, the ideal planner was conceived as a 'generalist' – someone who had the skill of bringing together and 'synthesising' all the considerations relevant to the development of a town, plus the skill of 'integrating' all these considerations together creatively into a single plan (see Cherry, 1974, p. 165). Because of this, as Eric Reade (1987) was later to point out, it was extremely difficult (if not impossible) to monitor the effectiveness of each of the proposals contained within a plan.

It is sometimes said that town planning has now lost its earlier vision and idealism. Certainly, the post-war period in Britain was a time of optimism, but, as we shall see in Chapters 3 and 5, it is questionable whether this Utopian comprehensiveness brought about better urban environments. It is therefore also questionable whether the passing of this Utopian tradition in town planning is a matter for regret.

Anti-urban aestheticism

In introducing this chapter, I described how the normative political context of British post-war planning brought together both radical and conservative elements, and that the combination of ideas from the political left and right laid the foundations for the post-war social democratic consensus. I also described how British post-war planning embraced the modernist project of turning its back on tradition and reconstructing the world anew, and that this was seen, for example, in the predisposition towards *comprehensive* planning and redevelopment, as well as the project of building *new* towns. But British planning did not entirely turn its back on tradition. Alongside the tendency towards radical modernism sat a strongly conservative set of values which emphasised protection, conservation, containment, – in other words, a general resistance to certain aspects of modernising change.

This conservatism emerges clearly in Gracey and Hall's account of the social values that underlay British post-war planning, which emphasised the maintenance of social stability and harmony, and careful stewardship of the land (Hall *et al.*, 1973, Vol. 2). Gracey and Hall (*ibid.*, p. 367) acknowledge that this position did not entail opposition to all change, but there was nevertheless 'a strong desire to control the effects of change, directing it in ways which preserve the traditional values of society'. In Hall's account (*ibid.*, Chap. 1) of the objectives of post-war planning, he sees this conservatism operating at three spatial scales: at the *national-regional* scale of the country as a whole; at the scale of *subregions or city regions*; and at the *urban* scale (the last scale of objectives is discussed more fully below).

At the *national-regional* scale the overall objective was to maintain the existing balance of population and employment between the main regions of Britain. Regional policy discouraged excessive economic growth in the already relatively prosperous regions of the south east and the west Midlands and, conversely, encouraged economic development in those regions in the north of England, Scotland and Wales that had suffered most heavily from economic depression and unemployment in the 1930s.

The objectives of planning at *subregional* or *city regional scale* can be summed up broadly under the label of 'urban containment', for all the objectives Hall describes at this level worked towards that end. There was a general aim to restrict or 'contain' the further growth of urban areas, partly because it was believed that huge urban agglomerations did not constitute ideal urban settlements and partly to protect the countryside from uncontrolled urban sprawl. Urban sprawl, in particular, had been a source of major concern in the interwar years as low-density suburban development spread out from all of Britain's cities (the Restriction of Ribbon Development Act 1935 was an early response to this problem). These twin concerns of urban containment and rural preservation were addressed simultaneously by establishing green belts around major cities (*ibid.*, Vol. 2, pp. 52–9). As it was appreciated that it was not possible to stop further urban growth completely, attempts were made to control and direct this. Scattered development (e.g. blocks of new development attached to already existing settlements, infill development, etc.) were discouraged; strong service centres (e.g. 'key settlement' policy in rural areas) were encouraged, as were (especially in new towns) self-contained and balanced communities (*ibid.*, pp. 38–64). Taken as a whole, the planning objectives Hall describes sought to maintain and reinforce the existing, inherited pattern and form of urban settlement throughout the country.

However, a deeper conservative value lay at the heart of British town planning thought – a general stance of 'anti-urbanism'.[6] Two complementary tendencies, in particular, constituted this anti-urbanism. On the one hand, there was the desire to preserve and maintain, wherever possible, the countryside and traditional rural settlements. On the other was a dislike of large cities, especially the industrial cities which had grown up in the last century. As Rosemary Mellor (1977, p. 141) has written of the ideas of three of the leading pioneers of town planning of the early twentieth century – Howard, Geddes and Unwin –

> The common denominator of these pioneer figures is their whole-hearted rejection of industrial-urban society as it had developed. This rejection, coupled with their love of the countryside, led them all to advocate a new form of urban living based on the house and the garden, the neighbourhood and the small town.

Certainly, Ebenezer Howard wanted to combine the advantages of rural *and* urban life – he advocated garden *cities*. But, as Mellor indicates, Howard's ideal city was a small country town, not a busy metropolis. This point is well made by Elizabeth Wilson (1991, p. 101):

> Howard was not an outright anti-urbanist – he praised the civic virtues and the cultural benefits of cities – but he rejected the extremes and intensity of the huge metropolis, and believed that cities should never grow beyond a certain size . . . Garden cities were intended to be both more orderly and more aesthetic than the 'teeming metropolis'. Garden cities never teemed.

Wilson touches on two central reasons why Howard, and many other British planning theorists, preferred the countryside and smallish country towns to the 'teeming' metropolis. First, large cities (and especially large industrial cities) were seen to be places of actual or potential social disorder, including crime and insurrection. Secondly, large cities were widely regarded as ugly places, which was reflected in the dominant position occupied by the pastoral tradition in British art, in which the countryside, not the city, was the place on earth closest to paradise.[7] There was therefore a social and an aesthetic dimension to British anti-urbanism. As Peter Hall (in Hall *et al.*, 1973, Vol. 2, p. 51) wrote of the objectives of post-war British planning: '. . . these objectives were not economic . . . The objective was *social*: the preservation of a traditional way of life . . . A subsidiary argument was . . . *aesthetic*: there was a bonus in the preservation of a traditional countryside, for the benefit of town dweller and country dweller alike'.

The ordered view of urban structure

Hall (in Hall *et al.*, 1973, Vol. 2, p. 64) suggests that there were two basic objectives to post-war planning at the urban scale: a desire to improve the quality of the physical environment of urban areas and a desire to improve accessibility within towns. These twin concerns and the tensions between them emerge clearly in the famous Buchanan report, *Traffic in Towns* (Buchanan *et al.*, 1963). One of the main concerns of town planning had always been to plan for the efficient movement of vehicles and pedestrians around cities, and this in fact was the chief contribution civil engineers made to town planning. By the late 1950s the increasing use of motor vehicles in Britain's cities was creating serious traffic congestion, and it was against this background that, in 1960, the government commissioned a study under the chairmanship of Colin Buchanan into the effects of motor traffic on British cities. *Traffic in Towns* was the result of this study.

The opening chapter of the report claims that the main challenge facing town planning is to provide easy access for motor vehicles whilst at the same time maintaining the quality of the towns' environments. But maximising accessibility with motor vehicles and maintaining environmental quality conflict with each other, and the proposals in the report to resolve this conflict tell us a great deal about the concept of the ideal urban structure current at the time.

One response might have been to recommend a move towards a much more dispersed, low-density pattern of urban development, so as to spread out traffic movements and their attendant problems more thinly (perhaps along the lines of Frank Lloyd-Wright's modern urban Utopia, 'Broadacre City'). After all, the Buchanan report (1964 edn, p. 38) itself admitted that: 'The conflict between towns and traffic obviously stems from the physical structure of towns. The manner in which the buildings and streets are put together is basically unsuitable for the motor vehicle.'

However, the Buchanan report rejected this option in favour of maintaining the traditional, compact structure of British cities. As the report (*ibid.*, p. 40) put it:

> For the period ahead into which we may reasonably peer – probably extending a little beyond the end of the century – it is reasonable to suppose that towns and cities will continue to exist broadly in their present form, for in spite of their acknowledged defects they contain great accumulations of material and cultural wealth'.

This stance in favour of maintaining the traditional compact form of urban settlements dovetailed neatly with the objective of protecting the countryside by containing the outward growth of urban areas. The report (*ibid.*, pp. 40–1) endorsed the 'anti-urban' preference for the countryside as a central justification for maintaining compact cities:

> . . . although persuasive arguments can be adduced in favour of urban dispersal, this island is not big enough for large-scale dispersal if a sensible relationship is to be maintained between developed areas and open country. Opinions may vary as to the future role of agriculture, but no one could question the special importance of the countryside to the people of this crowded island, nor would anyone be prepared . . . to see a large part of it sacrificed to a major dispersal of urban areas in order to accommodate motor traffic. After all, to 'get out into the countryside' is one of the main reasons why people buy cars'.

Note how the prime reason given here for protecting the countryside is aesthetic; the countryside is not to be 'sacrificed' to urban development because, presumably, the countryside is typically an aesthetically higher-quality environment than the town. But in addition to these 'negative' arguments against urban dispersal, the report (*ibid.*, p. 41) also claimed that traditional compact urban settlements were preferable in their own right:

> There are long-standing, well-tried advantages in the principle of compactness for urban areas which are not to be lightly jettisoned in favour of the supposed advantages of dispersal. In a compact area, journey distances, including the all-important journeys to work and school, are kept to a minimum. The concentration of people makes it possible to provide a diversity of services, interests, and contacts. There is a wider choice of housing, employment, schools, shops, and recreational and cultural pursuits. It is easier in a compact society to maintain the secondary activities, such as restaurants, specialist shops, and service industries, which all too easily fail if there is not a large clientele close at hand. The issue is not starkly between high density flats and low density suburbs – towns should have both – but whether to maintain or abandon the degree of compactness and proximity which seems to contribute so much to the variety and richness of urban life.

With our present concern with environmentally sustainable development, these arguments in favour of the compact city might be viewed as prescient but they are in fact significant for what they tell us about the prevailing values of British planning thought at the time. The sentiments expressed are intuitive judgements based on the kinds of urban environments people in Britain were used to. And what emerges is a conservative preference for what has been inherited from the past, the traditional ('long-standing, well-tried') compact form of towns. This is why a city like Los Angeles, whose dispersed low-density pattern reflected a new urban form that accommodated the motor car, was considered in British town planning circles at the time as a negation of everything a city should be.

Given the decision to accept the traditional form of the city, the report's two proposals for resolving the conflict between accessibility and environment followed logically. First, cities were to be organised into a patchwork of discrete 'environmental' areas whose environmental quality would be protected by permitting only local traffic, on 'local distributor' roads, into them. A network of main traffic-carrying roads would run between these areas. The main roads would ensure accessibility to all parts of a city by car, but since these roads would not penetrate local areas this high level of accessibility would be achieved at the same time as protecting local environmental quality (Figure 2.3). In this way, the report believed, transport accessibility and a high-quality urban environment could be reconciled.[8]

Buchanan's proposals presupposed that the well planned city was composed of an orderly 'cellular' structure of geographically distinct neighbourhoods or 'environmental areas'. This presupposition was central to the view most town planners held at the time of the ideal urban structure. As Lewis Keeble (1969, p. 10) had put it: 'The town ought to have a clear legible structure.' This ordered view of the ideal city found expression in two other ways.

First, the current orthodoxy was that the major land uses of the city should be clearly distinguished and provided for in separate 'zones'. Again, we see this

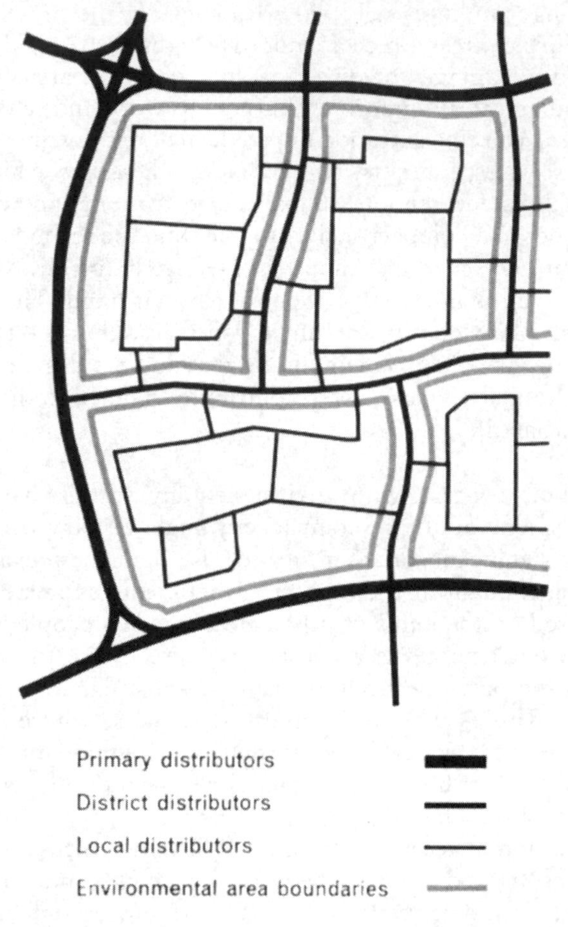

Figure 2.3 Planning for motor traffic and the environment at the same time: Buchanan's proposals for a hierarchy of roads and environmental areas
Source: From Buchanan *et al.*, 1963, Figure 13

ordering principle in most of the influential Utopian schemes for ideal cities, such as Le Corbusier's 'radiant city'. This principle found expression in, for example, the plans for the Mark 1 new towns, or in the zoning of London's Southbank as a single-use 'cultural zone' containing the Royal Festival Hall, the National Theatre, the Hayward Gallery, etc. This 'zoning mentality' came in part from a justified desire to separate obnoxious activities, such as heavy industries, from residential areas but, as London's Southbank shows, by the middle of the twentieth century the tidy-minded separation of different land uses had become an accepted norm among planners.

Second, the ordered view of urban structure found expression in spatially distinct 'neighbourhoods' conceived as village-like communities. This idea originated in the 1920s in the work of the American sociologist Clarence Perry, who recommended the division of the city into distinct 'neighbourhood units'. Each neighbourhood was to be provided with its 'own' local communal facilities, such as convenience shops, a local park, a church and a primary school, all located at the centre of the neighbourhood so that they would be within walking distance of, and act as a social focus for, the residents of the neighbourhood (Perry, 1939). In this way each neighbourhood would be relatively 'self-contained', rather like a traditional village in the countryside.[9] As has already been seen, this 'cellular' concept was evident in the master plans for all the post-war 'Mark 1' new towns and was also central to Abercrombie's 1944 plan for Greater London, in which the different districts of London (Islington, Stepney, Poplar, Hampstead, etc.) were to be 'reorganised . . . as separate and definite entities . . . with their own shops and schools' (quoted in Alexander, 1965).[10]

Christopher Alexander (1965) later described this ordered way of structuring the city around neighbourhood units as a 'tree' structure, because the parts of the city (the neighbourhoods) are linked to the centre but are otherwise viewed as relatively independent of each other. According to Alexander, this ordered way of thinking characterised nearly every plan made by town planners and designers in the first half of the twentieth century.

Implicit in this notion of the ideal urban structure were two value-laden presuppositions. First, the primary organising unit is the local neighbourhood which, when planned to be self-contained, has its own community life and autonomy within the larger whole, like a village within the city. Even large metropolitan cities of world significance, such as London, were seen as a patchwork of village-like communities rather than as bustling centres of commerce and culture with a busy interchange between their component parts. The focus was on the small-scale parts and the locality, not on the larger whole and the functioning of a great metropolis. This image is one we have met before in the garden city, an image that looked back not just to a traditional urban form but beyond to a simpler rural past in which communities were indeed villages. Again we are back with the anti-urbanism inherited from Victorian romanticism and medievalism. As Donald Foley (1960, cited in Faludi, 1973a, pp. 82–3) put it:

> The social ideology that emerged is essentially this: the best community life is provided in small, reasonably low-density communities. Building upon the traditional form and social organisation of the village, an image of desirable community life is held up as an ideal . . . The notions of the small community and of the small dwelling with garden seem to reflect a British value on smallness and a corresponding suspicion of large size. Running throughout the British social ideology of cities upon which town planning had drawn is the distinct and strong suspicion that great cities do not provide really decent living places.

The second value in the 'tree-structured' vision of the ideal city is the value accorded to orderliness itself – with its neatness, tidiness and the clear separation of things one from another. It is the vision which, as well as sorting the city into separate 'self-contained' neighbourhoods, also insisted on segregating different land uses into their own distinct zones.

An assumed consensus over the aims of planning

In his account of the ideology of British town planning, Donald Foley (1960, cited in Faludi, 1973a, p. 78) suggests that the ideal of basing the structure of cities on physically discrete neighbourhood units with low-density housing provided planning with a social-welfare-enhancing mission, in that this ideal constituted 'a broader social programme . . . providing the physical basis for better urban community life.' However, even if this is true, the traditionalist nature of the normative ideals of British planning ensured that this 'social programme' was barely radical. Thus although town planning (like 'planning' generally) was associated in many people's minds with left-wing politics, the objectives planners sought in the post-war years appealed to the political right as well as the left. As we have noted before, this was a time of broad consensus politics over the role of the state in society.

Post-war planning theorists also assumed that there was a consensus in society over the values and ideals town planning should embody. At least there was little sign amongst town planning theorists that these values and ideals were in any way controversial. Rather, the 'principles' of good planning were seen as self-evident – matters requiring 'common sense' rather than critical intellectual inquiry. The very use of the word *Principles* in Lewis Keeble's (1952) book is significant here. Keeble did not consider the principles he set out as matters of great ethical or political dispute. It was simply a case of setting out what were 'obviously' sound principles of good planning, and then putting these into practice.[11]

Two important points emerge from this. First, because consensus was assumed, it was further assumed that the ideals which planning sought were, uncontentiously, in the 'public interest'. As several writers have pointed out, planners in the post-war era assumed a 'unitary' view of the public interest (see, e.g. Meyerson and Banfield, 1955, p. 289; Taylor, 1994b in Thomas, 1994, pp. 106–7). This unitary view was given perfect expression by Lord Reith (1946, cited in Hall *et al*., 1973, Vol. 2, p. 70) when advocating the new towns programme in Britain: 'There are few in this country who would increase existing conurbations; there are few in this country who would not feel that the suburban sprawl of the past hundred years is deplorable from every point of view.' There is in this statement the ease of one who confidently believes that he knows what 'most people in the country' want, and that what they want is to live in small towns close to the countryside – in 'garden cities', in fact. The public preference for this traditional (even preindustrial) kind of settlement is assumed to be self-evident.

Secondly, given this assumed consensus, the task facing town planners was simply a practical one, of finding the 'technical' means to achieve given objectives, not debating these objectives themselves. And the appropriate means to achieve these planning objectives were the plans themselves – blueprint plans, in fact, – specifying the future pattern of urban development which would most closely approximate to the idealised vision of urban settlement described in this chapter. This emphasis on the technical and 'practical' problems of planning dovetailed with the conception of town planning which I described in Chapter 1; town planning was viewed as primarily an exercise in physical planning and design, and, given agreement on what kind of settlements should be planned for, the task was the practical one of planning and designing them. This 'technicalist' view of town planning also supported the view held by the Town Planning Institute that town planning was a distinct 'profession', involving a special, technical expertise, namely, a skill in the physical planning and design of towns.

Because the aims which town planners at this time sought were not regarded as especially controversial, and because, in virtue of this, planning was seen as largely a technical or practical exercise, the aims of town planning were not viewed as being *politically* contentious, and so town planning was not regarded as being an inherently 'political' activity. It was acknowledged that town planning operated within a political *context*, in which there was a requirement for development plans and decisions controlling development to be approved by an elected local authority and/or central government; and in which there was a comprehensive system of town and country planning that had been established after the Second World War. But given this, since the principles of good town planning were assumed to be self-evident and agreed by all, the *task* of town planning was not seen as politically contentious.

CONCLUSION

In Chapter 1 I described the conception of town and country planning which reigned in Britain when the new '1947 system' of planning came into being. I described how post-war town planners typically *defined* their activity, and I suggested that they saw town planning as essentially an exercise in the physical planning and design of land-use and built form. In this chapter I have gone on to describe the normative ideals which drove this physical planning in post-war Britain.

Given the prevailing conception of town planning as centred on physical planning and design, it was natural that the normative ideals which town planning sought – its vision of a better urban future – were conceived in 'physicalist' terms; in terms of the overall pattern and size of urban settlements across the country as a whole, and the more detailed spatial structure and layout of the different constituent parts of urban settlements. This is precisely what we find. Thus the ideal urban future is envisaged in terms of where, and how big, cities should be; in terms of a certain balance between city and countryside; and in terms of a vision of how, ideally, cities should be internally structured.

But even when town planning is conceived in terms of physical layout and design, there still remains room for radically different visions of the ideal urban future. And if we look at the particular normative vision which prevailed in post-war Britain, we find that, with the exception of the acceptance of design ideas drawn from modern architecture (and notably the ideas of Le Corbusier) which were applied in 'high-rise' council housing schemes throughout the country, the prevailing vision was one which looked back to an imagined, idealised past. It was a vision which saw the countryside as more beautiful than the town (certainly than the towns produced by the industrial age), and which regarded the best kind of town as about the size of a small country town or, if this was impossible to attain, then a city which was composed of discrete 'neighbourhood' areas with which people could identify, like villages within the city. The idea that the complex teeming metropolis itself might be a desirable living environment did not seriously come into the picture. The values of British town planning thought and practice in the immediate post-war years were thus deeply anti-urban, and in this respect, conservative.

NOTES

1. The main difference would have been over the financial provisions for taxing betterment which Labour attached to the Town and Country Planning Act 1947, and which the Conservatives dismantled when they were returned to power in the 1950s.
2. For discussions of the Uthwatt Committee's report, and of the land values issue generally in relation to planning at this time, see Cullingworth, 1975, Vol. 1; Mckay and Cox, 1979, Chap. 3; Cox, 1984, Chaps. 3 and 4.
3. The British '1947 system' of land-use planning has thus been appropriately described as a 'regulatory' system by, e.g. Mckay and Cox, 1979, Chap. 2.
4. A third influential vision of the ideal future city was that put forward by Frank Lloyd Wright, in his scheme for 'Broadacre City' (see Fishman, 1977). However, with its vision of sprawling low-density cities, this version of an urban Utopia received little serious attention in Britain, mainly because of the land-take that the realisation of such cities would involve. Los Angeles was hardly the ideal Frank Lloyd Wright had in mind, but the low-density 'spread-outness' of Los Angeles exhibited some of the features of Wright's 'Broadacre City'. And in British planning circles Los Angeles was generally regarded through the 1950s and 1960s as the *opposite* of the ideal city, and thus the kind of city which town planning should be seeking to *prevent*. Reyner Banham's book on Los Angeles, published in 1971, was the first architectural text to counter this negative view of Los Angeles, and it is only relatively recently, amongst postmodern urbanists, that Los Angeles (and so perhaps, by implication, Wright's 'Broadacre City') has been held up to represent (in parts at least) a contemporary ideal (see, e.g. Soja, 1995).
5. The rejection of tradition by modernist architects like Le Corbusier was mirrored by the modern movement in the other arts. Thus cubist, surrealist and especially abstract painting broke with the traditional idea of painting as a representational art; 'stream-of-consciousness' writing, like that in the last part of James Joyce's *Ulysses*, paid no heed to literary conventions of paragraphing and punctuation; atonal music disregarded the norm of composing music in a certain key; and so on.
6. For discussions of anti-urbanism in British planning see, e.g. Glass, 1959; Foley, 1960; Williams, 1973; Mellor, 1977, Chap. 4; 1982; Wilson, 1991, Chap. 7.

7. Wordsworth's 'Lines Composed a Few Miles above Tintern Abbey' is perhaps the literary exemplar of this ruralist tradition. More generally, the British romantic movement of the nineteenth century was, as much as anything, a movement inspired by the vision of a rural idyll. On this see, e.g. Clark, 1969, Chap. 11; Williams, 1973.
8. The Buchanan 'solution' to the problem of planning for high levels of motor traffic had been first suggested by Alker Tripp in the 1940s (see Tripp, 1942).
9. The 'Radburn' principle of designing residential development so that the pedestrian and vehicular circulation systems were separated was widely accepted as part of good neighbourhood planning. This idea originated in the USA from the work of Henry Wright and Clarence Stein (see Stein, 1958).
10. In Bell and Thywitt (1972, p. 407).
11. Interestingly, there was remarkably little attention given in Keeble's book to the problems of putting his principles into practice – to the issue of 'implementation' as it is now called.

3

Early critiques of post-war planning theory

INTRODUCTION: THE 'GOLDEN AGE' OF POST-WAR PLANNING?

The historian Eric Hobsbawm (1994) has described the period from the end of the Second World War to the early 1970s in western Europe and North America as a 'golden age' in what has otherwise been a bleak and unstable century. In spite of an initial period of post-war austerity (it was not until 1949 that rationing was abolished in Britain) and the emergence of a 'cold war' (with the threat of nuclear war) between the USSR and the capitalist west, the post-war period was one of renewed hope and optimism. This optimism was reinforced in the 1950s and 1960s when the economies of the capitalist west experienced high growth rates combined with full employment. Many people began to enjoy material standards of living which had hitherto been unknown, giving rise to talk of the 'affluent society' (Galbraith, 1958). The general sense that these were good times was expressed by Prime Minister Harold Macmillan. When boasting of his government's record prior to the British general election of 1959, he told the British people 'You've never had it so good'.

The period from 1945 to the late 1960s was also a kind of golden age for British town planning. As I noted in Chapter 2, a broad 'social democratic' consensus reigned in politics under which both major political parties endorsed an enhanced role for the state in managing society, including town planning.[1] The amount of physical development which took place added to the role town planning had to play. The first priority was the reconstruction of those parts of Britain's cities which had been bombed during the war. The need to manage reconstruction properly during the war itself was the impetus for establishing a stronger system of development planning after the war. There was also a perceived need to replace the huge legacy of Victorian 'slum' housing which remained in Britain's inner cities and to build new housing for the increasing population generated by the wartime 'baby-boom'. The post-war Mark 1 new towns had to be built and, on top of all this, there were calls to build new schools, hospitals, shopping centres and so on as part of the brave new world that was to be created after the deprivations of two world wars and the economic depression of the interwar years. Finally, rising standards of living in the 1950s inevitably brought new demands for development. Amongst these was the demand for new roads as more people began to acquire motor cars and towns became increasingly clogged with traffic. The 1963 Buchanan report (*Traffic in*

Towns, described in Chapter 2), was a manifestation of this concern. In short, the twenty-year period following the Second World War saw a great deal of physical development, and all this was to be overseen and planned by the newly established post-war planning system.

With hindsight it was perhaps inevitable that many mistakes would be made in post-war reconstruction and development – that, in Lionel Esher's (1981) phrase, the wave of post-war optimism would turn out to be a 'broken' one. By the end of the 1950s the planning that had 'created' this development was subject to increasing criticism, directed initially at the *practice* of town planning. Implicit in this criticism was a critique of the *theory* of town planning which underpinned this practice, and it is with these 'early' criticisms of planning theory that this chapter is concerned. My aim is to provide a broad summary of the kinds of theoretical criticisms which emerged in the late 1950s and early 1960s of the physicalist conception of town planning. In the third section of this chapter, I describe the criticisms which emerged at this time of the normative theory of post-war planning. In section 4, I briefly examine some of the concerns that came to be expressed in official circles in the early 1960s about the operation of the '1947 system' of planning. Specifically, I examine the criticisms of the 1947 development plan system which were made by the Planning Advisory Group established by the government in 1964. As we shall see, some of these 'official' criticisms mirrored the theoretical criticisms contained in the academic literature.

Before this, however, a general point needs to be borne in mind. All the criticisms of planning thought and practice (i.e. the institutionalised practice of town planning carried out by local planning authorities) assumed that planning was largely responsible for the development which appeared on the ground – so that if this development was considered 'bad' then this was the result of bad planning. As sociologists Ray Pahl and others (Pahl, 1970) have put it, a 'managerialist' view of planning was assumed, in which town planners (together with other local authority officials) were the 'managers' of urban areas and so responsible for the quality of the development. This view has since been contested, since statutory planning is only one agent amongst many shaping the pattern and form of physical development; planning and planners cannot be held entirely responsible for development outcomes. Even in the post-war heyday of public sector activity, the prime determinant of physical development was 'the operation of market forces subject to very little constraint' by planners (Pickvance, 1977, in Paris, 1982, p. 69). However, this perspective on the power of planning only really emerged in the 1970s, and is examined in Chapter 6.

CRITICISMS OF THE PHYSICAL AND DESIGN BIAS OF TOWN PLANNING

Two levels of criticism of post-war urban development emerged in the 1950s: criticisms of the *quality of the design* of new development, and of the *emphasis on physical planning*. Criticisms of the quality – and especially the aesthetic quality –

of post-war architecture were particularly pronounced. Two writers in particular spear-headed these attacks: J.M. Richards and Ian Nairn. In the late 1940s and early 1950s, Richards (1950) criticised the tame new neo-Georgian office blocks which appeared in London and other provincial cities (such as the rebuilt city centre of Exeter). In a special edition of the *Architectural Review* entitled 'Outrage', Nairn (1955) launched a vigorous attack on the undistinguished 'neither-town-nor-country' character of early post-war suburban housing developments, which he dismissed as 'subtopian'. In these ecriticisms was an implicit attack on the practice of design control, although some of this criticism was insufficiently focused (again, a crude managerialist view of local authority responsibility was too readily assumed).

The development plan system created by the Planning Advisory Group is discussed later in this chapter; here the focus is on the criticisms that emerged of the emphasis town planners placed on physical planning and design, which struck at the very roots of the whole concept of post-war town planning.

Criticisms of social blindness

In 1957 a book was published describing early post-war redevelopment in the inner London area of Bethnal Green written not by an orthodox *planning* theorist but by two sociologists, Michael Young and Peter Willmott. The first part of the book was an account of the traditional working-class community in Bethnal Green, with its close ties of family and kin. The book went on to describe what happened to this community when, following the redevelopment of a large part of Bethnal Green, many of its members were moved out to a new suburban housing estate on the edge of London, Greenleigh.[2] The tale is a sad one. The migrants were given better houses, with gardens, and new schools and shops and, from interviews following their move, Young and Willmott found that many of the relocated residents felt themselves to be living in a better *physical* environment. What is sad, however, was the loss of the *social* environment – the community – they had known in Bethnal Green. As one woman put it: 'When I first came I cried for weeks, it was so lonely' (*ibid.*, p. 122).

Young and Willmott were aware that this loss of social contact was not entirely attributable to physical dislocation: as people's material standards of living improved through the 1950s, new social (or rather 'anti-social') habits were forming, such as watching television. These drew more and more people into the private sphere of their homes and away from external 'public' activities that formerly brought them into contact with other people. However, the physical move to Greenleigh still played a significant role in the loss of community life, and this was in part a result of town planning.

Although some Bethnal Greeners had moved to Greenleigh of their own volition (to be near family and friends), by far the majority had been moved as a result of the London County Council's housing policy (*ibid.*, p. 127). Young and Willmott (*ibid.*, p. 198) therefore concluded that, where housing redevelopment occurs it would be better to do this as far as possible 'by moving as a block the social groups, above all the wider families, to which people wish to belong'. This it was a simple suggestion: why hadn't planners and housing managers thought of it before?

Young and Wilmott suggest that that town planners had hitherto overlooked the social aspects of housing redevelopment because their attention was focused primarily, or even solely, on physical matters, such as the age and condition of people's homes. It was not that town planners did not care about people but that they were wedded to a view of town planning which concentrated on the physical environment. Thus in places like Bethnal Green the planners saw a slum because, physically, this was a slum.[3] They failed to notice, however, that, socially, Bethnal Green was anything but a slum – it was a healthy close-knit community. The conception of town planning as physical planning led planners to attend only to the physical environment, and hence to ignore or overlook the non-physical social environment in which people lived.

This is not to say that town planners did not think about 'social' matters. There was much talk, for example, about planning 'neighbourhoods' and planning for community life (as noted in Chapter 2). The problem was, however, that, even when planners attended to such matters, they tended to view them in terms of the physical environment. As Young and Willmott (*ibid.*) put it:

> The physical size of reconstruction is so great that the authorities have been understandably intent upon bricks and mortar. Their negative task is to demolish slums which fall below the most elementary standards of hygiene, their positive one to build new houses and new towns cleaner and more spacious than the old. Yet even when town planners have set themselves to create new communities anew as well as houses, they have still put their faith in buildings, sometimes speaking as though all that is necessary for neighbourliness was a neighbourhood unit, for community spirit a community centre.

What is being criticised here is not just a certain set of planning practices but the theory which underlies that practice. Because they were bound to an essentially physicalist conception of town planning, planners tended to view towns and their problems only in physical (and aesthetic) terms. Because of this they simply *did not pay attention to social matters*; their theory of planning prevented them from really *seeing* social issues. This is borne out in the surveys town planners undertook in preparation for their planning. In areas like Bethnal Green copious surveys were done of the 'age and condition' of buildings and of whether, physically, houses were 'fit for human habitation', as well as of the general aesthetic character of such areas. But *social* surveys, of the kind Young and Willmott undertook, were not done. Which was why the social consequences of physical redevelopment were not foreseen.

Criticisms of physical determinism

Planners did not ignore social considerations entirely; there was a concern to plan for community life by planning neighbourhoods. However, in planning neighbourhoods, town planners tended to assume that neighbourhoods in the social sense could be created by planning neighbourhoods physically with 'neighbourhood shops', a 'local' primary school and so on. In other words, an assumption was made that the layout and form of the physical environment would shape, even 'determine', the quality of social life. Such a belief has appropriately been dubbed 'physical determinism' (or 'architectural' or 'environmental' determinism). This was well described by Maurice Broady (1968, pp. 13–14):[4]

> The theory has been expressed as follows: 'The architect who builds a house or designs a site plan, who decides where the roads will and will not go, and who decides which directions the houses will face and how close together they will be, also is, to a large extent, deciding the pattern of social life among the people who will live in these houses.' It asserts that architectural design has a direct and determinate effect on the way people behave. It implies a one-way process in which the physical environment is the independent, and human behaviour the dependent variable. It suggests that those human beings for whom architects and planners create their designs are simply moulded by the environment which is provided for them.

Sociologists such as Maurice Broady (*ibid.*, p. 15) further criticised this theory, arguing that the primary factors determining community life were social, not physical:

> Of much more importance in explaining neighbourliness are the *social* facts, first, that the people who lived in the slums had often lived in the same street for several generations and thus had long-standing contacts with their neighbours and kin; and second, that people who suffer economic hardship are prone to band together for mutual help and protection. It is true that neighbourliness is induced by environmental factors. Of these, however, the most relevant are social and economic rather than physical.

Sociologist Ruth Glass was also critical of the idea that social neighbourhoods could be created by physical planning. In the 1940s she had undertaken a study of working-class districts in Middlesbrough (Glass, 1948) and found that the geographical areas within which people lived were much more complex and overlapping than the planning idea of neat, physically distinct (and 'relatively self- contained') neighbourhood areas. In other words, the real-life networks of social activities and relationships were not simply contained within clearly bounded geographical areas.

To sum up the two criticisms so far described of the physicalist bias of post-war town planning theory, the theory was criticised at two levels. At one level it was criticised for concentrating on the physical environment to the extent of *ignoring* the social environment. And at another level, to the extent that town planners did consider the social environment in their plan making (e.g. in planning for 'neighbourhoods'), they were criticised for assuming that the shape of the physical environment *determined* the social environment.

Lack of consultation

In the immediate post-war era town planners tended to assume they knew best what sorts of physical environments were unfit for people to live in. Accordingly, they did not even *consult* the inhabitants about how they would like to see their surroundings planned. This point was made forcibly in Norman Dennis's (1970) study of comprehensive redevelopment in Sunderland. Here, just as in Bethnal Green, huge swathes of Victorian working-class housing were scheduled for clearance and 'comprehensive redevelopment'. As Dennis points out, within some of the designated clearance areas there were pockets of housing which were in sound condition and so perfectly fit for human habitation, or capable of being made so with some basic renovation. Even in terms of the local council's own physically based criteria, housing was earmarked for demolition which need not have been. Many people living in these clearance areas also wanted to remain in their old homes, even when they were considered inadequate in terms of the council's criteria of 'fitness'. *Some* people wanted to be rehoused but others did not, as Dennis (*ibid*., p. 333) found out after consulting the local communities: 'Four out of every ten families living in the 1965–70 clearance areas were very satisfied with their present living conditions; in some areas the proportion was as high as six out of every ten . . . Two out of every three owner-occupiers were not in favour of demolition.'

Yet demolition went ahead. This was not because the local authority deliberately went against people's wishes, for such a view presupposes a cynical interpretation of planners' motives. It was simply that the planners did not know what the wishes of the local community were because they did not consult them. Although this was partly a result of their physicalist conception of planning, it was also because professional judgements about what constituted a good living environment were assumed to be uncontentious. When it did come to the planners' attention that their proposals did not meet with people's wishes, it was assumed that the planners' judgements were more sound (they were the 'experts' after all), and that ordinary people did not have such a clear perception of what was in their best interests. As the Labour Minister for Town and Country Planning, Lewis Silkin, said: 'I think it is necessary to lead the citizen – guide him. The citizen does not always know exactly what is best' (quoted in Ward, 1994, p. 112).

This lack of consultation with the people whose environments were being ripped apart exhibits the planners' failure to appreciate an elementary theoretical distinction between matters of fact and matters of value. The judgements made were assumed to be purely technical 'professional' judgements, and hence planners did not think it necessary to consult residents' views or, if objections were drawn to their attention, they assumed they knew best. What planners failed to appreciate was that a judgement about what constitutes a worthwhile living environment is a value judgement, not one of pure 'technical' fact. Therefore, planners were further criticised for not recognising the value-laden and therefore 'political' nature of town planning. This became an important criticism of prevailing planning theory and practice in its own right (quite apart from its association with the physicalist bias of planning) and we return to it again later in this chapter.

CRITICISMS OF BLUEPRINT PLANNING

Although the level of detail and precision in the first generation of British post-war development plans (or 'town maps') did not match that of architects' or engineers' finished drawings, that model was imitated to the extent that the boundaries of land-use zonings were precisely delineated and, within zones, details of the density of development and plot ratios were also specified. However, there was a problem with this. Whereas a building would typically be constructed from architects' drawings over the period of a year or so, the realisation of a town plan could take twenty or thirty years, or even longer. Furthermore, a town's development did not stop there; once a plan was 'completed', development and redevelopment would still go on, and so a new plan would be needed to guide this later development. Planning a town was thus an ongoing continuous process in a way that designing a building was not.[5]

This ongoing nature of town planning was acknowledged under the 1947 Act by the requirement of local planning authorities to review their development plans every five years. In this respect at least, town plans were not seen as 'end-state' documents. In another sense, however, they were, for in making detailed prescriptions for how each site within a town was to be used, including detailed prescriptions for densities of new development, a plan put forward a fixed view of the future. This level of precision failed to take seriously two very important facts.

First, detailed site-specific land-use zoning did not allow for the possibility that, over time, changes could occur in the processes affecting urban development. For example, a successful industry might want to expand on to a neighbouring site or an unsuccessful one contract and sell off some of its land. Likewise, a successful row of shops might attract new retail development and a school with a 'baby-boom' catchment area might require extra space for new buildings or playing fields and so on. The point was well made in a paper by Maurice Brown on 'Urban form' (1966). Brown drew attention to the tension which inevitably existed between, on the one hand, any plan for a town which attempted to specify its future form by 'freezing' it into fixed land-use zones and, on the other, the fact that in any functioning settlement land-use activities were in a constant state of flux (see also Brown, 1968).

Brown illustrated his point hypothetically. He analysed the case of a town plan where, in order to relieve pressure on the town centre, equal amounts of land for five inner-suburban shopping centres around the main commercial core of the town were zoned. As Brown pointed out, it is likely that each centre would be developed at a different time, and that some centres will become better established and so more successful than others, thereby generating demands for more space than the uniform amounts allocated for each centre (Brown, 1966, p. 7). The first-generation development plans deployed at this time were in fact applied during a period of accelerating change, resulting in precisely the kind of fluid picture Brown described theoretically in his article. Brown raised serious questions about the appropriateness of detailed blueprint or 'master' plans as vehicles for planning a town's development (*ibid.*, p. 3):

> Every plan in the course of implementation is liable to come up against unforeseen occurrences or accidents. As an instrument of public policy it must have the capacity to survive such accidents with only minor modifications. If it is mutilated beyond recognition or even finally abandoned, besides defeating its own objects, it brings into disrepute the whole concept of planning as a reliable instrument of public policy.

Brown was aware that his argument could suggest abandoning the blueprint model and that such a suggestion might be met with incredulity: 'The notion of doing away with master plans may seem at first a little curious. It is almost unthinkable that any new town or any new city could be built without one' (*ibid.* p. 8). Although Brown shied away from advancing a specific proposal as to how plans might be better conceived, he did hint at an alternative model: 'we might see planning in the light of a game of chess, divided into a series of moves each limited and decisive in its own terms but each striving to secure maximum freedom for successful manoeuvre in the subsequent stages' (*ibid.* p. 9). Perhaps for the first time we have here the suggestion that town planning would be better conceived as an ongoing *process* rather than as a 'one-off' exercise in detailed site planning and design. In consequence, plans might be better conceived as flexible strategies.

Brown's paper was initially delivered to a meeting of the Town Planning Institute in London in 1966 at a time when an emerging body of theoretical work stressed the dynamic functioning of cities by viewing them (and the environment generally) as 'systems' of inter-related activities. Brian McLoughlin (1965a; 1965b) and George Chadwick (1966) had both published papers in the *Journal of the Town Planning Institute* advocating a 'systems view' of planning (see Chapter 4). Both these authors' arguments echoed Brown's in stressing the dynamism of urban functions and in calling into question the appropriateness of 'end-state' blueprint plans.

Secondly, although there was provision for altering the detailed zoning prescriptions of any plan through its five-yearly review, perfectly satisfactory buildings might be blighted by the detailed zonings, only for it to be realised later on review that this blight need never have occurred because the offending zonings are changed. This is precisely what happened under the detailed 1947-style development plans. For example, many development plans included road widening lines which sliced through the fronts of rows of houses, hence blighting them (i.e. selling them off and leaving them vacant to deteriorate). It was then often subsequently realised that such road widening proposals were no longer appropriate (e.g. because it was later discovered that what was needed were entirely new roads at other locations). Decent houses had thus been needlessly blighted and their inhabitants' lives needlessly disrupted. If these criticisms were true, a further implication was that it might be a mistake to think of town planning as being like architecture or, indeed, like any other physical design profession.

CRITICISMS OF THE NORMATIVE IDEALS AND ASSUMPTIONS OF POST-WAR PLANNING THEORY

In Chapter 2 I summarised the basic values of British post-war planning thought under four heads. I suggested, firstly, that ideas about the kinds of urban environments town planning should realise were often cast in terms of Utopian and comprehensive visions of ideal settlements; secondly, that the stated ideals of town planning exhibited an overriding concern with aesthetics, often of a conservative anti-urban kind; thirdly, that conceptions of ideal settlements exhibited an ordered, even 'tidy-minded' view of urban structure; and fourthly, that a 'technicalist' view of town planning predominated which assumed that there was a general consensus over what constituted the ideal urban environment and hence too over the values and aims of town planning.

By the mid-1960s all these hitherto accepted 'principles' of good planning (see Chapter 2) were being questioned seriously, and a more general and fundamental criticism was being voiced – that town planners lacked an adequate understanding of the phenomena with which they were dealing. It was alleged that the ideas underlying town planning showed a lack of understanding of how towns and cities actually functioned and, as a result, planners (and planning) were in danger of destroying the richness and vitality of cities and urban living. The fracturing of inner-city working-class communities, such as those in Sunderland or Bethnal Green, by comprehensive redevelopment exhibited this. The allegation was that town planning was driven by normative thinking that was grounded in very little empirical analysis and understanding of how towns and cities actually worked.

Criticisms of Utopianism

Unquestionably the most trenchant and compelling critique of town planning orthodoxy at this time came from the American writer, Jane Jacobs, in her book *The Death and Life of Great American Cities*, first published in 1961. Though grounded in the American experience of post-war city planning, what Jacobs said was relevant to the British experience. Indeed, her book remains arguably the most important planning theory text published since the end of the Second World War. Her most fundamental contention was that town planners typically exhibited very little understanding of the cities they were 'treating' because they had been preoccupied with simplistic Utopian visions instead of trying to understand and address the problems of real-life cities:

> Cities are an immense laboratory of trial and error, failure and success, in city building and city design. This is the laboratory in which city planning should have been learning and forming and testing its theories. Instead the practitioners and teachers of this discipline (if such it can be called) have ignored the study of success and failure in real life, have been incurious

> about the reasons for unexpected success, and are guided instead by principles derived from the behaviour and appearance of . . . suburbs, tuberculosis sanatoria, fairs, and imaginary dream cities – from anything but cities themselves. (Jacobs, 1961, p. 16)

Elsewhere (*ibid*, p. 23) she says: 'The pseudo-science of city planning and its companion, the art of civic design, have not yet broken with the specious comfort of wishes, familiar superstitions, over-simplifications, and symbols, and have not embarked upon the adventure of probing the real world.' Le Corbusier's 'radiant city' was a perfect example of this (*ibid*., p. 33):

> Le Corbusier's dream city has had an immense impact on our cities. It was hailed deliriously by architects, and has gradually been embodied in scores of projects, ranging from low-income public housing to office building projects . . . His city was like a wonderful mechanical toy . . . his conception, as an architectural work, had a dazzling clarity, simplicity, and harmony. It was so orderly, so visible, so easy to understand. It said everything in a flash, like a good advertisement . . . But as to how the city works, it tells . . . nothing but lies.

Criticisms of anti-urbanism

The lack of understanding of real-life cities was also evident in planners' anti-urbanism and preference for a tidy, ordered view of urban structure. Jacobs was critical of both these aspects of normative planning theory. According to her, one reason why town planning theorists were incurious about real-life cities was because they had typically rejected the large metropolitan city as a desirable settlement form. Even Le Corbusier's 'radiant city', in spite of its pretensions to be a plan for a 'great city', was in reality a very anti-urban construct with its vast open spaces and acres of 'grass, grass, grass' (Jacobs, 1961, p. 32).[6] Apart from Le Corbusier's Utopia, the most influential alternative vision of the ideal city derived from Howard's garden city, which in America was taken up and developed by a number of influential 'decentrists', including Lewis Mumford, Clarence Stein, Henry Wright and Catherine Bauer. According to Jacobs, all these leading planning theorists showed little interest in examining how large cities worked because their vision of the ideal settlement was deeply anti-urban and so excluded the large city. As Jacobs (*ibid*., pp. 27–9) says of Howard:

> The programme he proposed . . . was to halt the growth of London and . . . repopulate the countryside . . . His aim was the creation of self-sufficient small towns, really very nice towns if you were docile and had no plans of your own . . . Howard set spinning powerful and city-destroying ideas: he conceived that the way to deal with the city's functions was to sort and sift out of the whole certain simple uses, and to arrange each of these in relative self-containment . . . He was uninterested in the aspects of the city which could not be abstracted to serve his Utopia.

> In particular, he simply wrote off the intricate, many-faceted, cultural life of the metropolis. He was uninterested in such problems as the way great cities police themselves, or exchange ideas, or operate politically, or invent new economic arrangements, and he was oblivious to devising ways to strengthen these functions because, after all, he was not designing for this kind of life in any case.

Criticisms of the ordered view of urban structure

Jacobs was also critical of the neatly ordered view of the ideal urban structure, with its prescriptions for tidying up land uses into separate zones; its correspondingly tidy conception of structuring cities in terms of a cellular pattern of distinct neighbourhoods; its neat separation of pedestrian and vehicle traffic in 'Radburn' layouts (even in relatively quiet residential areas); and so on. Jacobs criticised most of these standard principles by simply turning them on their heads and showing that the converse of what was conventionally recommended is often just as desirable.

Take, for example, the ideal of planning urban land uses into distinct homogeneous zones. According to Jacobs, it is the *mixture* of uses, not their tidy separation, which is a precondition of good city life. For a mixture of uses generates more activity throughout the day and night and so adds to the diversity and vitality of an area. She is therefore scathing about planners' adherence to 'the principles of sorting out – and of bringing order by repression' as in, for example, the 'idea of sorting out certain cultural and public functions and decontaminating their relationship with the workaday city' (Jacobs, 1961, p. 35). Instead, she proposes (*ibid.*, pp. 23–4) an alternative principle: 'One principle emerges so ubiquitously, and . . . becomes the heart of my argument. This . . . is the need of cities for a most intricate and close-grained diversity of uses that give each other constant mutual support, both economically and socially.'

Jacobs' espousal of complexity and diversity was echoed in another important critique of town planning thought, which appeared in 1965 – Christopher Alexander's article 'A city is not a tree'. In this, Alexander contrasts cities which have grown more or less spontaneously over a long time – which he calls 'natural' cities – with cities which have been deliberately created by designers and planners – which he calls 'artificial' cities. One might take issue with Alexander's terminology here, since all cities are human-made and therefore, in this sense, 'artificial'.[7] Nevertheless, the point of Alexander's distinction is to draw a contrast between the 'ordering principles' of cities which have grown in a largely piecemeal (and in this sense 'natural') fashion over a long time, and those cities (or parts of cities) which have been completely, or comprehensively, planned and built 'in one go' – such as the British post-war new towns, Le Corbusier's Chandigarh in the Punjab, or Levittown in the USA.

Given this distinction, Alexander goes on to make two claims. First, comrehensively planned 'artificial' cities lack some 'essential ingredient' which 'natural' cities possess, and that it is this ingredient which makes natural cities

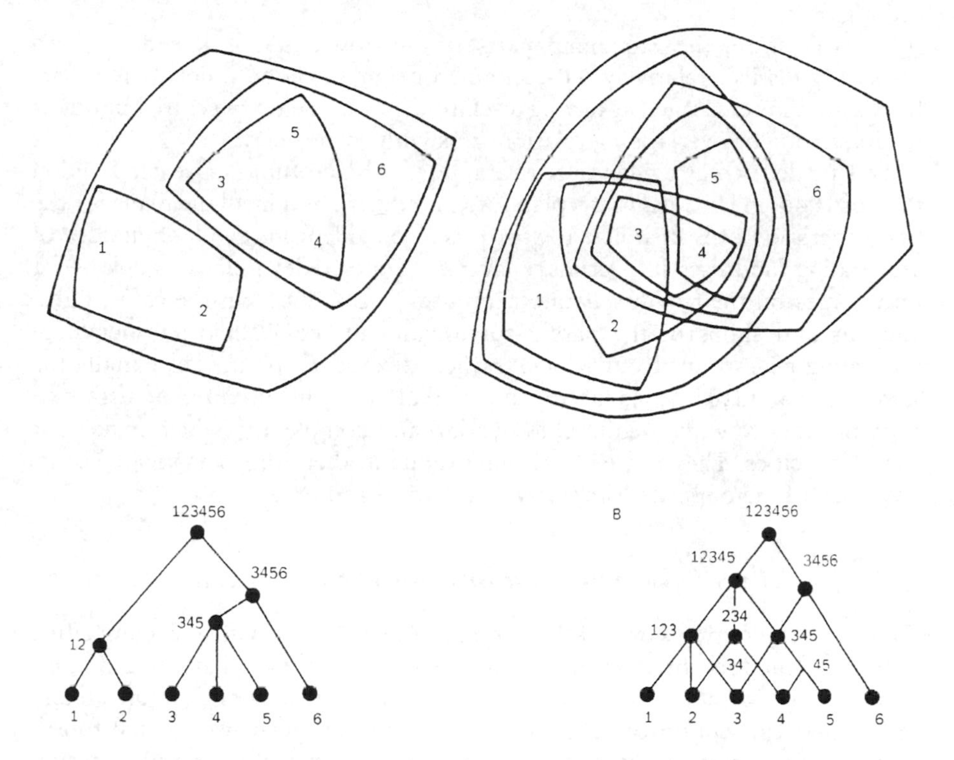

Figure 3.1 A tree structure
Source: Alexander, 1965

Figure 3.2 A semi-lattice structure
Source: Alexander, 1965

more interesting and successful places than modern planned cities. The second is an attempt to explain this missing ingredient in 'artificial' cities in terms of the 'ordering principles' which govern them as compared with 'natural' cities. Alexander uses some rather off-putting jargon to describe this, but basically his point is that comprehensively planned 'artificial' cities are based on an over-simplified conception of urban form, which he terms a 'tree' structure. A tree structure is one in which the various parts of the whole exist as separate entities. These parts have a relationship to the whole structure but they do not overlap with, or relate to, each other. If we think of such a structure in a hierarchichal form then it would look like a tree, with separate branching parts, as shown in Figure 3.1. However, some structures are more complex than this because the parts we can distinguish not only relate to the whole structure but they also relate to, or overlap with, each other. This more complex kind of structure Alexander calls a 'semi-lattice' (Figure 3.2).

All this is put abstractly, but Alexander's thesis is that so many modern planned cities are sterile places to be in – lacking the richness and interest of long-standing 'natural' cities – because they do not have that essential ingredient of complex overlapping relationships which are the hallmark of successful cities. Alexander cites the way that most modern town plans prescribe a neatly ordered cellular structure for a town (a tree structure) in which, around the town centre, the remaining parts of the town are broken down into physically distinct relatively self-contained neighbourhood units. Neighbourhoods are thus envisaged as having a relationship to the town centre but not as having strong or overlapping relationships with each other.

This tendency was evident, for example, in Abercrombie's Greater London Plan of 1944, in Le Corbusier's plan for Chandigarh and in all the plans for the first generation of British new towns. It is also evident in the planning idea of segregating land uses into primary land-use zones (Alexander cites plans for universities on their own separate campuses – isolated from the cafés, pubs, cinemas and shops of the parent town), and in the Radburn principle of separating pedestrians from vehicle traffic. Alexander's point is essentially the same as that made by Jacobs, namely, that it is the *mixture* of uses and activities in a town, which implies overlap and complexity, which makes for successful cities. The failure of planning thought, therefore, was seen to be its advocacy of the opposite: simplicity, separation and order.

The critique of the consensus view of planning values

Taken together, the work of Jacobs and Alexander provides a compelling critique of the first three normative principles of post-war planning thought I described in Chapter 2: its Utopian comprehensiveness, its anti-urbanism and its ordered conception of urban structure. The main criticism of the fourth principle – a consensus view of planning values – came from another quarter. We have already noted how traditional planning thought and practice, with its emphasis on the physical environment, was sociologically ill-informed and naive about local urban communities. It was also sociologically naive in its more general assumptions about society as a whole. Even in the 1960s, when there was much talk about planning being 'for people', there remained a tendency to speak of 'people' or 'the public' as an undifferentiated group. This is, of course, far from the truth: the public of any modern society is composed of all sorts of different groups, with differing and sometimes incompatible interests. This was much stressed by sociologists in the 1950s and 1960s (e.g. Coser, 1956; Dahrendorf, 1969; Rex, 1970), and some social theorists proposed a 'conflict' model of society as an alternative to the consensus model assumed by many planning theorists.

That there were different groups with different interests relevant to physical planning policy had been exposed in America in the mid-1950s by Meyerson and Banfield (1955) in their study of housing policy in Chicago. Meyerson and Banfield showed that groups in different positions (e.g. with respect to income) had quite different views about what the priorities of public housing policy should be. This lack of consensus, therefore, made the business of deciding what principles to enshrine in housing policies and plans a highly contentious 'political' exercise. As Meyerson and Banfield (*ibid*., p. 316) put it:

> if only one end were relevant in the making of a plan, it would be a simple matter for the planner to choose a course of action. But almost always there

> are numerous ends to be served and no one course of action will maximise the attainment of all of them.

The same point was stressed by another American, Melvin Webber (1969, p. 286):

> It is fruitless – and certainly misleading – to compute overall community values. In a complex urban society there is no viable single community. And, because each of a multiplicity of competing communities values things against different value scales, there can be no set of generalised values or criteria against which to appraise a project. There can only be a plurality of competing values held by a plurality of affected groups.

If town planning thought and practice was in these ways sociologically naive, then it was politically naive too. In Chapter 2 we saw how the assumption that there was a ready consensus over the ends which planning should pursue went hand-in-hand with a 'technical-professional' model of town planning. If the values and principles of good town planning were self-evident and generally agreed, there was little need for the public, or their political representatives, to participate in debating town planning matters. But once it was acknowledged that there was no necessary consensus over the ends which planning should pursue, and that, indeed, different groups in the public might *disagree and dispute* what these ends should be, then town planning came to be seen as a potentially controversial 'political' activity. From this perspective, the view that town plans and planning decisions were just technical matters, best made by expert professionals, also seemed too simple.

It is not surprising, therefore, that, by the 1960s, planners were being criticised for failing to involve the relevant public in discussions of their plans or for failing, even, to take account of expressed public attitudes to planning proposals. Dennis's studies (1970; 1972) of housing redevelopment in Sunderland have already been discussed in this respect. Another study in Newcastle upon Tyne by Jon Gower Davies (1972) was similarly critical of 'evangelistic bureaucrats'. More specifically, Davies claimed that his study showed how decision-making in planning reflected and so tended to reinforce further the inequalities between richer and poorer groups in society. For the power to make or influence decisions over such matters as redevelopment 'lies where it has always lain: with the possessors of large amounts of wealth, power and influence', so that planning becomes 'highly regressive . . . with those who have least suffering most, and those who have a lot being given more' (*ibid*., p. 228).

Traditional town planning theory, therefore, was accused of failing to appreciate the differential distributive effects of planning action on various social groups holding different, and sometimes conflicting, values and interests.

OFFICIAL REFLECTIONS: THE REPORT OF THE PLANNING ADVISORY GROUP

The criticisms described in this chapter came mostly from 'academic' writers and scholars. However, criticisms were also voiced in official circles in the report of the Planning Advisory Group set up by government to review the operation of the British post-war development planning system (*The Future of Development Plans* – the 'PAG report' – Ministry of Housing and Local Government, 1965).

The recommendations of this report proposed two different kinds of development plan which were subsequently adopted in the Town and Country Planning Act 1968. This Act placed an obligation on local planning authorities to introduce a 'two-tier' system of development plans: broader-level strategic or 'structure' plans and, nesting within these, more detailed district or 'local' plans. This distinction between 'strategic' and 'local' plans has remained in force ever since, but the PAG report was also important for its criticisms of town planning theory as described in this chapter. The authors make it clear that, in reviewing the 1947 planning system, their concern was not with 'planning policies as such' but rather with 'the broad structure of the planning system' and hence with development plans, which 'are the key feature of the system' (*ibid.*, p. 1). This is important because its criticisms were directed at the form of plans, and hence at the conception of planning that underpinned these plans and not at the normative ends which, it was assumed at this time, town planning should realise.

Three key criticisms of the 1947-style development plans stand out. First, plans were criticised for being too detailed and therefore *inflexible*. In setting down precisely delineated, site-specific land-use zonings, development plans were ill-suited to accommodate unanticipated changes affecting land use and physical development and so soon became outdated. The report (*ibid.*, para. 1.21, p. 5) claimed that the 'original intention' of the architects of the 1947 system was that the land-use allocations in development plans 'should be drawn in with a "broad brush" and that the rigidity of detailed zonings . . . should be avoided'. However:

> the statutory definition and the notational techniques adopted have resulted in a constant tendency towards greater detail and precision. The plans have thus acquired the appearance of certainty and stability which is misleading since . . . it is impossible to forecast every land requirement over many years ahead . . . [And so] it has proved extremely difficult to keep these plans not only up to date but forward looking and responsive to the demands of change. The result has been that they have tended to become out of date. (*Ibid.*, paras. 1.21, 1.23, pp. 5, 6)

This point is emphasised again later in the report: '[a] plan cannot be a static document because it is not dealing with a static situation' (*ibid.*, para. 6.11, p. 37). This is substantially the same criticism of the rigidity of the 'blueprint' model of planning voiced by Maurice Brown and others. And although ostensibly 'not about theory', in questioning the rigidity of a certain style of development plan the report was at the same time bringing into question a certain *theory* of planning – namely, one which conceived of town planning and its products as similar to the physical design professions, such as architecture.

The second main criticism concerns the *scope* of town plans as conceived under the 1947 Act. In some respects, this criticism was allied to the first: in failing as flexible documents, the old-style development plans 'have become more and more out of touch with emergent planning problems' (*ibid.*, para. 1.24, p. 6). Implicit in this criticism was the suggestion that town plans (and therefore town planning) should address a wider range of issues than those encompassed by planning which focused predominantly on matters of physical design and aesthetics: 'the 1947 style development plans deal inadequately with transport and the inter-relationship of traffic and land-use; . . . they fail to take account quickly enough of changes in population forecasts, traffic growth and other economic and social trends' (*ibid.* para. 1.23, p. 6).

This statement can be taken in two ways. At one level, it could imply that town planning, still conceived of as primarily concerned with matters of physical form and design, should ensure that plans are 'kept up to date' – for example, by making more flexible plans which can be changed to address forces affecting physical development (i.e. the first criticism described above). At another level it could be taken as implying that town planning should not be just confined to matters of physical form and design but should be concerned with 'economic and social' policy more generally, having a wider remit than that envisaged by town planners wedded to a physicalist and design-based conception of planning.

The third main criticism was, perhaps paradoxically, the opposite of the first – that, for some purposes, the 1947-style development plans were not specific enough: 'the present development plan system is too detailed for some purposes and not detailed enough for other purposes' (*ibid.*, para. 1.31, p. 8). In particular, for the purpose of bringing about good-quality urban design, a simple land-use plan was too blunt an instrument:

> the development plans have not provided an adequate instrument for detailed planning at the local level. While town maps may present a reasonably clear picture of land-use, they do not convey any impression of how the land will in fact be developed or redeveloped, or what other action may be taken in the area to change its character or improve the environment. They give little guidance to developers beyond the primary use zoning. They make no contribution to the quality of urban design or the quality of the environment. (*ibid.*, para, 1.28, p. 7)

If true, this last statement was an especially embarrassing indictment of the operation of the town and country planning system and of the plans made under this. One of the central tenets of the town planning movement was to bring about an improvement in the quality of the environments people inhabited; *that* indeed was the main reason for the traditional predominance of an urban-design-based concept of town planning, and yet it seemed that the planning system was failing at this level too.

According to the PAG report, these problems arose from a failure to distinguish, at the level of development planning, between strategic planning at broad spatial scales and for long periods of time into the future, and more detailed planning for specific localities and for shorter time horizons. The problem with the 1947-style development plans was that they fell between both stools. Because of their detail they were not appropriate instruments for long-term strategic planning but because they were only land-use plans they were not 'fine-grained' enough to ensure good-quality site planning and urban design.

It is easy to see the logic of the report's proposals for two different kinds of development plan. At broader spatial scales and over longer time horizons, the report proposed 'broad-brush' strategic plans – 'structure' plans – which would show development proposals in only very general terms. Such plans could be cast in the form of a map (or a 'plan' in this sense), but equally they might be in the form of a series of policy statements rather in the manner of other forms of aspatial policy-making. Either way, such plans would be more flexible, capable of being updated and revised as appropriate. In this way, structure plans would overcome the criticisms made of the rigidity of detailed 'blueprint' plan created for the future. If such plans also encompassed a wider remit of 'social and economic' as well as just 'physical' issues, then they might also overcome the criticisms made of traditional town planning for its physicalist and design bias.

At the more detailed level of 'local' planning, where development activity is planned for in the short term, the report proposed 'local' plans. These were really the same as the old-style site-specific plans, but with the suggestion that they might contain even more detailed guidance to ensure higher-quality urban design.

CONCLUSION

In this chapter I have described the main criticisms of post-war town planning thought which emerged in the late 1950s and early 1960s. A common theme which unites these criticisms was the accusation that planners were insufficiently informed about the nature of the reality they were tampering with. For example, in relation to the comprehensive redevelopment of large areas of cities, and the physicalist and design-based conception of planning which underpinned this, what emerged was planners' lack of understanding of the 'communities' they were planning for – a lack of understanding which was illustrated the world over at this time, in Le Corbusier's Chandigarh in the Punjab just as much as in British local authority planning in Sunderland or Bethnal Green.

To the extent that planners did consider 'social planning', for example in planning neighbourhoods, they showed, in their assumptions about environmental determinism, an inadequate comprehension of the complex relationships between physical environments and social life. In the very plans which planners made, in seeking to set down master plans and blueprints for the future form of cities, planners failed to recognise the changing nature of cities.

Planners' lack of understanding of cities was also exhibited in their normative ideals at this time. In its Utopianism, its anti-urbanism, its simple 'tree-like' models of urban structure, and in its assumptions about consensus over what the ideals of 'good' planning should be, traditional town planning thought failed to grasp the complexity and richness, as well as the undoubted problems, of human social life and its manifestations in cities.

What planners lacked, and what planning theory had failed to provide, was an adequate *empirical understanding* of the world they were seeking to manipulate. More than anything, this explained the failures of planning in practice in the two decades following the Second World War, and it also explained the deficiencies in the planning theories which guided this practice. Any town planning activity, then as now, is based on value-judgements about how the world, or some part of the world, might be improved and made a better place to inhabit. Urban planning has to be guided by some *normative* ideals and principles. However, if such normative thinking is not grounded in an understanding of the real world then there is the danger of it being wildly unrealistic or just wrong. If planners seek to rehouse a community without first of all seeking to understand what sort of a community they are rehousing, or without being aware that there *is* a 'community' in an area to which development proposals apply, then we are likely to end up with the disasters of post-war comprehensive redevelopment. The critique of planning theory for its lack of empirical understanding was therefore a serious one.

The criticisms of planning thought and practice described in this chapter were taken seriously by some planning theorists. How they responded to these criticisms, and what new theories developed, is the subject of the next chapter.

NOTES

1. Although he was a Conservative Prime Minister, prior to the 1959 general election Macmillan boasted that his administration would build more *council* houses than Labour.
2. The very name of the new estate – Greenleigh – is significant in itself, for it reflects those anti-urban values described in Chapter 2.
3. In the light of the high-rise housing estates which replaced the old terraced streets in places like Bethnal Green, even this came to be contested.
4. The environmental determinism of this time is vividly captured in a statement by the American writer, Lewis Mumford, on cities and planning. On a visit to Britain he advised planners 'to plan for man as a human animal, to give him houses, neighbourhoods and towns which will teach him lessons of integrity and continuity, so that as he grows and matures, he will eventually go forth into and govern the world as a whole' (cited in Cherry, 1974, p. 140).
5. In fact, planning buildings and bridges is also an ongoing process in the sense that these structures need to be *maintained* through time. Furthermore, changes in function and lifestyles can soon make the fixed form of a building redundant. Hence the 1960s view that architects, as well as town planners, should incorporate flexibility into their designs and move away to some extent from designing very rigid 'blueprints'.
6. Le Corbusier estimated that only 5% of the land in his 'radiant city' would be built on. The remaining 95% would be park. And so 'the city shall become a park' (Le Corbusier, 1933, p. 205).
7. Many of the successful cities Alexander would have described as 'natural' had also been consciously planned. He is unspecific on his history here, but he is in line with many critics of town planning who have made the common assumption that most great cities were unplanned, and that it is the absence of planning which largely explains why such cities are attractive but for criticisms of this view see, e.g. Rorig (1932/55, Chap. 8), Gutkind (1969, Chap. 3), Taylor (1994a, pp. 36–7).

PART II
PLANNING THEORY IN THE 1960s

4

The systems and rational process views of planning

INTRODUCTION: THE RADICAL CHANGE IN TOWN PLANNING THOUGHT

In 1969 a fourth edition of Lewis Keeble's book, *Principles and Practice of Town and Country Planning*, was published. This was still substantially the same text as had originally been published in 1952. In the same year, 1969, another book was published which, like Keeble's book before it, was to become a standard text for planning students: Brian McLoughlin's *Urban and Regional Planning: A Systems Approach*. Figure 4.1 shows the covers of the 1969 edition of Keeble's book and McLoughlin's book, side by side. Nothing more vividly captures the radical change in town planning thought which took place in the 1960s than the contrasting images on these two covers.

On the cover of Keeble's book is one of the author's designs for a hypothetical town centre, an image which shows town planning as an exercise in physical planning and town design. Though hypothetical, it conforms to established conventions of what a plan for a real town centre might be like; indeed, the diagram could have been the plan of an existing town. The diagram on McLoughlin's book is obviously more abstract. None the less, this also purports to be a representation of a possible town or city in which the circles and triangles represent activities (or land uses) at particular locations, and the lines represent connections between these activities. The varying thicknesses of these lines represent different degrees of interconnectedness (e.g. in terms of flows of goods, or people, or traffic, etc.). In other words, the diagram on McLoughlin's book represents an image of the city as an active functioning thing – as a 'system'.

Why is the cover of McLoughlin's book like this? It is because McLoughlin puts forward the view that town planning is an exercise in the analysis and control of urban areas and regions *viewed as systems*. So if McLoughlin had been called upon to define town planning, he would have defined it is an exercise in *systems analysis and control*. This way of conceiving planning was a far cry from the post-war Keeblean view of physical planning as an exercise in design.

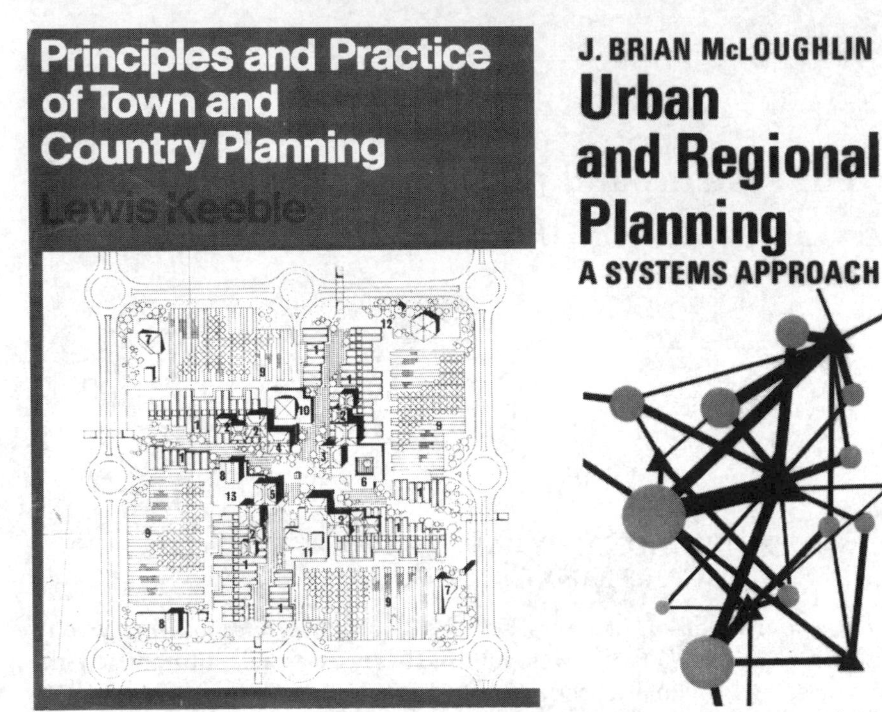

Figure 4.1 The covers of Lewis Keeble's *Principles and Practice of Town and Country Planning* (1969 edition) and Brian McLoughlin's *Urban and Regional Planning: A Systems Approach* (1969)

During the mid to late 1960s, two distinct theories emerged which are not (and were not in the 1960s) always clearly distinguished from each other. One was the 'systems view' of planning noted above, which was essentially derived from a theory of the *object* that town planning seeks to plan, namely, the environment (towns, cities, regions, etc.), now seen as a system of interconnected parts. The other was the 'rational process' view of planning, which was a theory about the *process* of planning and, in particular, of planning as a rational process of decision-making.

Both theories presumed a 'deep' conception of planning and control which sociologist Anthony Giddens (1994, p. 58) has described as a 'cybernetic' model of planning and politics, and indeed systems theory had direct links with what was called the 'science of cybernetics'. The idea of cybernetic control has also been associated by Giddens and others with modes of thinking and action characteristic of 'modernism'. I suggest that the systems and rational process theories of planning, taken together, represented a kind of high water-mark of modernist optimism in the post-war era, and in this they shared something in common with the earlier post-war view of planning, in spite of the other differences between these two periods of planning thought.

In both this and the following chapter I describe what I term 'new planning ideas *of the 1960s*'. This designation isn't arbitrary; it is fair to say that, in Britain, it was in the 1960s that the ideas I discuss here first had a significant impact on town planning thought. However, these ideas did not just emerge from nowhere, suddenly in the 1960s. Systems theory and rational decision-making evolved in other disciplines during the 1940s and 1950s, and they were 'imported' into town planning in the 1960s. Arguably, it was in the first half of the 1970s that these ideas had their widest influence on planning thought. Thus it was not until 1971 that a 'companion' volume to McLoughlin's book was published in Britain, George Chadwick's *A Systems View of Planning*. And it was in 1973 that, again arguably, the leading theorist in Britain of the rational process view of planning, Andreas Faludi, published his influential books *Planning Theory* and *A Reader in Planning Theory*.

THE SYSTEMS VIEW OF PLANNING

Basic concepts and their application to planning

The systems view of planning was frequently described in highly abstract, technical and mathematical terms, but the basic ideas of systems theory are really very simple. At the heart of general systems theory is, obviously, the idea of things as 'systems'. A 'system' is something composed of interconnected parts. The *Oxford English Dictionary* defines a system as 'a complex whole, a set of connected things or parts', and also as 'a set or assemblage of things connected, associated, or inter-dependent, so as to form a complex unity'. There are two things to note here. First, any system has some kind of coherence or unity which enables us to distinguish it from other systems and so view it as an entity ('a complex whole') for the purposes of investigation. Thus a system is analogous to a 'set' in the way this term is used in mathematics, in which what is common to a set is what unifies it and at the same time distinguishes it from other sets. Secondly, what makes a system is not just a set of distinct parts but the fact that the parts are interconnected, and so interdependent. The structure of a system is therefore determined by the structure of its parts and their relationships (see again the diagram on the cover of McLoughlin's book in Figure 4.1).

The interconnections between the parts of a system are central to its functioning. Consider, for example, human beings as systems. Our bodies comprise various distinguishable parts: the heart, the lungs, the kidneys, the liver, the brain, etc. When functioning healthily, these parts are actively interconnected via the circulation system, which takes blood from the heart to the brain, liver and kidneys, etc. In the technical language of systems theory, one could say that a dead body is one in which the interconnections between the body's parts have 'ceased to function'.

All living organisms could be viewed as 'systems', but the use of the term needs to be qualified here, for any organism depends on, and is therefore related to, its environment. The whole of reality is one integrated system, and any system we distinguish within this, such as a living organism, is really a 'subsystem' within this larger whole. Because of this, the functioning of any system (or subsystem) has to be understood in terms of the ways its parts are 'externally' interconnected with 'parts' of other systems as well as 'internally' with each other. To take our example of human beings again, the healthy functioning of a person's lungs depends on the quality of the air breathed in from the 'external' environment or 'ecosystem', because any human being is a subsystem within a larger 'ecosystem'.

Just as we can think of living organisms as systems, so too we can view functioning human-made entities, such as cities and their regions, as systems. A city can be viewed as a system in which its parts are different land-use activities interconnected via transport and other communications media, i.e. a land-use/transport system. As the planning theorist Brian McLoughlin (1965a, p. 260) put it in his earliest article on the systems view of planning, 'The components of the [urban] system are land uses and locations which interact through and with the communications networks'. This only describes the objects town planning deals with, but the systems view of planning follows logically from this concept of the environment as a system. If the physical environments (towns, cities, regions, etc.) town planning seeks to plan and control are systems, it follows that town planning can be defined as a form of systems control. Or, to put this more fully, since exercising intelligent control over a system requires a prior understanding of the system to be controlled, then we can define town planning as a form of systems analysis and control.

This way of seeing town planning was not entirely new. In the early years of the twentieth century, the pioneering planner Patrick Geddes wrote of cities and their regions as functioning entities, analogous in this respect to living organisms (Geddes, 1915; interestingly, Geddes was trained as a botanist). However, apart from his strictures on the need to undertake surveys prior to planning (see below), Geddes's ideas remained marginal to the mainstream of town planning thought throughout the first half of the twentieth century, which continued to be dominated by architectural ideas. And so by the 1960s, against the background of a design-based view of planning, the systems view struck many planners as novel, even revolutionary. Below, I describe five major differences between the systems view and the traditional design-based view described in Chapter 1.

Once it was acknowledged that cities (or regions, etc.) were complex systems, it became all the more clear that planners needed to understand 'how cities worked'. Geddes had emphasised the importance of undertaking surveys prior to the preparation of plans, and his method of 'survey–analysis–plan' had been widely adopted. Yet town planners had not acquired a deep understanding of how cities actually functioned. As noted in Chapter 3, Jacobs's and Alexander's main criticism of traditional planning theory was that it showed a serious lack of understanding of the complex reality which planners were dealing with. Alexander drew attention to the fact that planners seemed to lack an appreciation of the complex and rich *inter-relationships* between phenomena which give rise to successful cities.

Secondly, once cities are viewed as *inter-related* systems of activities and places, it follows that a change to one part of the city will cause changes to some other part. Any proposed new development must be evaluated in terms of its probable effects, including its effects on activities and places far beyond the actual sites where the new development was proposed. For example, in considering a proposal for a major new shopping development on the edge of a town, planners should also examine the likely effects of this development on, for example, the town centre shops, the local transport system, the likelihood of other development pressures arising in its wake and then of the further effects of *that* new development and so on.

This was a significantly different way of examining and assessing development proposals from that which had been typically undertaken by planners who viewed planning largely in terms of design and aesthetics. It suggested the need for a new kind of planner altogether, one who was trained in analysing and understanding how cities and regions functioned spatially in economic and social terms – a planner, that is, trained in economic geography or the social sciences rather than architecture or surveying. Hence McLoughlin's suggestion that the appropriate theoretical understanding needed by town planners was to be found in 'location theory', and he devoted a whole chapter of his book to introduce this theory to planners (McLoughlin, 1969, Chap. 3).

Thirdly, as noted in Chapter 3, there were serious questions about whether it was appropriate to produce detailed 'end-state' master plans. Systems theory, with its emphasis on activity, dynamism and change, suggested the need for more adaptable flexible plans – such as the broad-brush 'structure' plans proposed in the PAG report. When McLoughlin described 'plan formulation', he envisaged town plans as 'trajectories', not end-state blueprints for a fixed future (*ibid.*, Chap. 9). As he put it: 'The form of the plan is that of a trajectory of states at suitable time-intervals' (*ibid.*, p. 255).

Fourthly, acceptance of urban change also suggested a view of town planning as an *ongoing process* of monitoring, analysing and intervening in fluid situations, rather than an exercise in producing 'once-and-for-all' blueprints for the ideal future form of a town or city.

Fifthly, viewing cities (or other discrete areas of the environment) as systems of interconnected activities implied considering them economically and socially, not just physically and aesthetically. This suggested a much broader and more ambitious remit for planning than previously (remember Keeble's dictum that town and country planning was *not* economic and social planning).

This conception of planning was illustrated in *The Spirit and Purpose of Planning* (Bruton, 1974) where chapters entitled 'Social planning' and 'Economic planning' sit alongside one entitled 'Physical planning'. This broader concept of planning was reflected in the new 'structure' plans introduced by the Town and Country Planning Act 1968. These plans were specifically intended to be *strategic* planning documents and their purposes were considered to be as much economic and social as physical. Planning was no longer defined as chiefly involving skills of *design* and physical planning (though both these more traditional concerns would have been acknowledged as having a more significant place in detailed local planning work and development control).

Related to this last point, in the late 1960s a gap developed between planning theory and the practice of town planning at the local level. No doubt this was attributable to the abstract, highly technical (and frankly abstruse) language of systems theory, with its talk of mathematical modelling, 'optimisation' and so on. The division also derived from the fact that much of the work of local authority planning offices continued to be at the level of 'local' planning and development control where a constant stream of applications for planning permission had to be dealt with. At this level matters of design and aesthetics continued to be regarded as central. Planning theory concerned with much broader systemic considerations tended to be seen as irrelevant by the everyday local planner with a heavy case-load.

It was therefore only in the newly emergent field of strategic, 'structure' planning introduced by the Town and Country Planning Act 1968, that some of the ideas associated with systems theory were considered by practitioners as relevant to their practice. Keeble's book had been much used in planning offices, and whatever one thinks of that book as a theoretical work, through it the worlds of planning theory and practice met. The same could not be said of McLoughlin's *Urban and Regional Planning: A Systems Approach*, still less of Chadwick's later *A Systems View of Planning*, in which certain passages were specially marked off for the 'faint-hearted' planner who was 'less than literate mathematically' (Chadwick, 1971, p. xii)!

The rise of the systems view of planning

The emergence of the systems view of planning could be explained as a response to the criticisms of the traditional 'physical design' view of town planning. The systems view of planning certainly seemed to meet three of those criticisms. In concentrating on the *physical and aesthetic* qualities of the environment, traditional town planning theory and practice exhibited a lack of understanding of the social and economic life of cities. With its aim of understanding how cities worked as activity systems, the systems view implied a commitment to understanding the social and economic life of cities. Secondly, traditional town planning theory exhibited a lack of understanding of the *complexity and inter-relatedness* of urban life. With its avowed aim of seeking to analyse and 'model' the complex inter-relationships of cities as systems, the systems view promised to meet this criticism. Finally, the traditional 'master' plans had been criticised for their inflexibility. With its stress on strategic and flexible plans which were sensitive to the dynamic, changing nature of cities, the systems view promised to overcome this problem too.

The emergence of the systems view of planning can thus be seen as a logical response to the deficiencies of the physicalist theory. However, it would be naive to suppose that developments in planning thought occur in this logical way or that they occur simply as a result of developments *within planning thought*. Wider forces were at work that contributed to the rise of this new theory. As the American historian of science Thomas Kuhn (1962) has shown, fundamental changes in thought (what he calls 'paradigm shifts') are not just driven by the accumulation of evidence and the rational response to this evidence. Wider technological, sociological and psychological factors also play a part.

The inter-relatedness of urban phenomena, and specifically land use and transport, had been widely recognised by transport planners in both the USA and the UK, and had already been highlighted by the 1963 Buchanan report. This, coupled with rising car use and traffic problems, generated calls for proper 'land use/transportation' studies. Transport planners in the 1960s were mostly traffic engineers, and mathematics was central to their training. Mathematically 'modelling' land-use/transport relationships and flows was thus something transport planners took up with enthusiasm. Systems theory had originated in the highly technical fields of operations research ('OR') and cybernetics where the precise modelling of systemic relationships using statistical and mathematical techniques was seen as necessary to control systems. The development of computers capable of handling the data derived from numerous systemic relationships also encouraged the use of systems theory in town and country planning.

Systems theory had an impact on a number of other disciplines. In academic geography, for example, the traditional concern with space and location was translated into viewing settlements and land uses as locations within networks of inter-related places – as spatial systems (see, e.g. Haggett, 1965, pp. 17–23). As the town planning profession had been opened up to graduates from disciplines other than architecture and surveying, etc., and as it was geographers who mostly exploited this opening, cross-fertilisation occurred between these two disciplines. As noted earlier, a whole chapter of McLoughlin's book (1969, Chap. 3) promoted geographical work on location theory rather than design theory, as the necessary theoretical foundation for planning.

The 'quantitative revolution' in 1960s geography was driven by a desire to make geographical studies more precise and 'scientific' rather than an 'art'. The impressively sounding language of systems theory and analysis and rigorous statistical methods of investigation promised to place geography on a much firmer theoretical and scientific foundation, thereby improving its standing as an academic discipline. The same could be said of planning, which had also been traditionally conceived as an art. The trenchant criticisms of writers such as Jane Jacobs seemed to show that the practice of planning lacked any firm theoretical foundation. Systems theory, with its claim to be 'scientific', seemed to offer this hope for planning, just as it did for geography.

Finally, the ecological thinking of the late 1960s emphasised the inter-relatedness of natural phenomena in 'ecosystems'. Again, the opening chapter of McLoughlin's book, introduced the reader to the basic ideas of systems thinking by describing 'man in his ecological setting', and by illustrating the practical relevance of understanding systemic relationships through examples of human actions which had irreparably damaged natural ecosystems.

So although the emergence of the systems view of planning in the 1960s can be explained in part as a rational response to the alleged deficiencies of the traditional design-based theory of planning, this is only part of the story. In the case of planning, following the damning criticisms which had been made of its practice and theory in the early 1960s, as much as anything there was a felt need for the discipline to acquire an intellectually firmer foundation. Systems theory, with its technical and seemingly sophisticated vocabulary of 'modelling', 'mathematics' and 'science', seemed to provide this. It was therefore not surprising that it was taken up enthusiastically by a generation of younger planners riding high on the optimism of the 1960s.

THE RATIONAL PROCESS VIEW OF PLANNING

Introduction

Although the systems view of planning derived from a concept of the environment as a system, what this view of planning did not address was the best method, or process, of *doing* planning. One of the first theorists to draw a distinction between the *process* of planning and the *object* or *substance* planning deals with was the American, Melvin Webber: 'I understand planning to be *a method for reaching decisions*, not a body of specific substantive goals . . . planning is a rather special way of deciding which specific goals are to be pursued and which specific actions are to be taken . . . the method is largely independent of the phenomena to be planned' (Webber, 1963, cited in Duhl, 1963, p. 320).

This distinction was also emphasised by Andreas Faludi in his book, *Planning Theory* (1973b, Chap. 1). Faludi similarly distinguishes between 'substantive' planning theories about the object (i.e. planning deals with the environment) and 'procedural' planning theories about the process or procedures of going about planning. Faludi also described substantive theories as theories *in* planning, and procedural theories as theories *of* planning. And since the latter were literally theories of *planning*, Faludi held that planning theory was, or should be, about procedural theory. Although this idea later attracted criticism (see Chapter 6), what is important to note here is the way the substantive/procedural distinction emphasises that the systems view of planning, based on a theory of the object (the 'substance') planners plan is, in Faludi's terms, a substantive theory, whereas the rational process view is clearly a procedural theory.[1]

It is worth recording that this was not the first theory about the process of planning which had been developed. The Geddesian dictum of 'survey–analysis–plan' has already been mentioned as a precursor to 1960s planning theory, and an analysis of this method may help to explain why the rational process view came to be preferred. The deficiencies of survey–analysis–plan (SAP) are threefold. First, to undertake a survey presupposes that there is some reason, or purpose, for carrying out the survey. In planning, the suggestion that one needs to undertake a survey to prepare a plan presupposes that there must be some problem one is trying to solve,and that it is the purpose of the survey to illuminate this. Logically, therefore, there is a planning 'stage' prior to carrying out the survey – namely, the definition of a problem (or an aim). For example, if we undertake a survey of traffic congestion prior to preparing a plan, this presupposes that traffic congestion is a problem; and the assumption that traffic congestion is a problem is made logically prior to any survey of it, and it is only in terms of this prior assumption that it 'makes sense' to survey traffic congestion.[2]

The second deficiency of SAP is that it implies, by the use of the word 'plan' in the singular, that there is only one possible plan, there may be alternative possible strategies. If there are, these alternatives need to be properly evaluated against each. To continue our example of traffic congestion, one plan for solving this problem might involve building new roads and/or widening existing ones. An alternative might involve retaining the existing network of roads and planning to relieve the congestion on them by encouraging the use of bicycles or public transport, etc. If, then, there are possible alternatives like this, it should be part of any rationally considered process of planning to set them out clearly, and then to evaluate them in terms of the problems they are designed to address. Obvious though this is, it is not made clear by the simple formula of 'survey–analysis–plan'.

Thirdly, in ending with the production of a plan SAP implies that the process of planning ends here. This, too, is misleading for we usually seek to implement the plans we make. In other words, a plan is usually followed by some *action*. 'Action' or 'implementation' is a further stage of the process of planning not mentioned by SAP. Furthermore, once implemented a plan or policy may turn out to be ineffective or it may have undesirable effects which we have not foreseen, etc. So it is also important to *monitor the outcomes* of our actions to check they are having the effects we want them to have and, if not, we may need to review and revise our actions or plans. To take our example of traffic congestion again, a plan to deal with this problem by building new roads may, when enacted, turn out to be fruitless because the new roads may attract more people to travel by car and so lead to a build up of congestion again. It is therefore crucial to 'monitor' the plans we adopt to see whether they are actually solving the problems we want to solve, but again, SAP does not acknowledge this.

From these criticisms we can see what a more considered account of a rational process of planning would have to include. If we visualise this process as involving (like the process of 'survey–analysis–plan') a number of distinct stages, then we arrive at something like the process shown in Figure 4.2. Figure 4.2 distinguishes five main 'stages' in a rational process of planning. First there must be some problem or goal which prompts the need for a plan of action. From an analysis of this, a definition of the problems or goals is arrived at. This analysis is necessary not only to guide any empirical investigation (or 'surveys') but also because, on closer inpection, the initial perceptions of the problems and/or goals may be questionable. It may be that the problem is not really a problem at all, or that what is a problem for one group may not be for another group, or that there are additional problems that were not noticed at first and so on.[3]

The second stage is to consider whether there are alternative ways of solving the problem (or achieving the aim) and, if there are, to clarify these. The third stage is to evaluate which of the feasible alternatives is most likely to achieve the desired end. In everyday life we are continually undertaking such evalua-

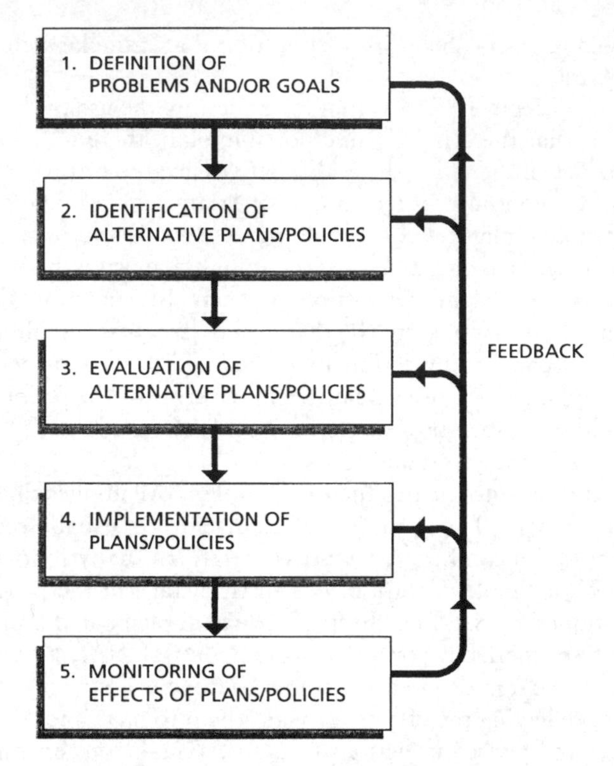

Figure 4.2 Planning as a process of rational action

tions when considering what best to do, and we generally do this 'intuitively'. But in complex decision-making situations the task of evaluating alternatives is, obviously, correspondingly complex, and may require a more systematic analysis of the likely consequences of implementing any alternative.[4] Sophisticated techniques, such as cost-benefit analysis, have therefore been developed for evaluating alternative options, and there is a large literature relevant to this stage of the planning process alone (see, e.g. Lichfield, Kettle and Whitbread, 1975).

The process of planning does not end when a decision has been made, for the chosen policy or plan then needs to be implemented. It is thus more accurate to describe the rational process of planning as a theory or model of rational *action*, rather than 'decision-making'. That is why Figure 4.2 shows 'implementation' as a further (fourth) stage in this process. There is yet a fifth stage which involves monitoring the effects of the plan to see whether it achieves the desired ends. A rational process of planning is thus an ongoing or continuous one. Rarely are our objectives achieved perfectly, and even if they are other objectives (or problems) invariably arise. Hence the feedback loops in Figure 4.2, indicating that a rational process of planning has no final end-state. Note how the feedback loops may return to *any* stage of the process: we may need to review our actions, revise our view of the problems, consider other alternatives which we didn't consider before, or accept our initial definition of the options but now question our original evaluation of them and so on.

The recognition that rational planning involved an ongoing, continuous process represented a significant break with the traditional design-based view of town planning. As noted in Chapter 1, town planners who saw their practice as essentially an exercise in physical design, tended also to see their task as one of producing blueprint plans for towns. In his book *Planning Theory*, Faludi (1973b, Chap. 7) drew a distinction between 'blueprint' and 'process' planning. The emergence, in the 1960s, of the view of town planning as an ongoing, continuous process (as well as a rational one) therefore implied a rejection of blueprint planning.

Sources of the rational process view

As with the systems view of planning, the rational process view originated from more general theory which had developed outside the field of town planning. In this case, the relevant theory was 'decision theory' (particularly general theory about rational decision-making) which was adopted and applied to town planning (see, e.g. Simon, 1945; 1960; Faludi, 1987, Chap. 2).

Two other factors provided a favourable context for the development of the rational process view. In government generally there was an increasing interest in corporate styles of management and decision-making. At the national level this was reflected in the social democratic style of politics which was characterised as a managerial politics in which technical professionals played a key role in advising politicians as to how best to manage the economy, the welfare state, etc. The 1960s was the high tide of the social democratic, 'corporatist' state, and the acceptance of rational planning was so widespread that some social theorists, such as Daniel Bell (1960), spoke of the 'end of ideology'. At local government level, this political stance was reflected in the adoption of strategic and corporate styles of management, following the recommendations of the Bains report (Bains, 1972) into the structure and management of local authorities.

Secondly, there was a renewed faith in the application of 'science' to policy making – not only in applying the findings of scientific research to policy but also in relation to the policy-making process itself (hence the talk of the 'policy sciences', 'scientific management', etc.). Although what made something 'scientific' was often not questioned, it was commonly assumed that the *quantification* of factors relevant to policy (such as traffic flows) was the hallmark of 'being scientific'; hence, if something could not be quantified (such as the beauty of a place) then it was not considered to be scientific (and hence often marginalised in policy-making).

More specifically, analogies were drawn between the scientific method as described by the philosopher Karl Popper (e.g. 1963) and the rational process of planning. Popper's account of scientific method insisted that any scientific inquiry was first of all driven by some belief or hypothesis about the world. The role of empirical investigation was then to test the veracity of the belief or hypothesis by – according to Popper – examining *critically* the belief or hypothesis (i.e. attempting to *falsify* it), so that only those hypotheses which withstood critical testing survived as credible theories. The rational process view of planning was sometimes likened to this. Definitions of planning problems or goals, or even plans and policies themselves, could thus be equated with scientific hypotheses that needed to be subjected to severe empirical (or 'scientific') testing before being implemented.

Further clarification of the rational process model of planning

On page 61, I described the basic model of rational decision-making and action which was adopted in the 1960s by those who took a rational process view of planning. Straightforward and uncontroversial though this model might seem, the idea of rational planning generated considerable debate amongst planning theorists. Some of these were critical of the rational process view, and some of their criticisms are examined in Chapters 5 and 6. Other planning theorists have sought to refine and develop the theory of rational planning, and some of this work is described below as a way of clarifying further the rational process view of planning.

Questions about rationality itself

In the simplest terms, a rational decision (or action) can be described as one for which *persuasive* reasons can be given. 'Persuasive' because 'reasons' of some kind can be given for doing almost anything, but having 'any' reason for doing something does not of itself make that action rational. A persuasive reason is one which connects directly with certain values we hold and aims we wish to attain. In so far as factual evidence is relevant to these values or aims, reasons are also persuasive if they are based on valid knowledge.

However, what constitutes a persuasive reason will often be contested. Sometimes there are clusters of persuasive reasons, some of which point in different directions, and so further judgements have to be made about the 'balance' of reasons. Moreover, as noted above, if some reasons are persuasive only in terms of certain values, individuals or groups adhering to different values may find different reasons persuasive. So adopting something like a rational process approach to planning cannot give us simple, formulaic answers to complex problems.

However, the rational process model of planning does imply an approach which gives reasons – and, it is hoped, persuasive ones – for the plans, policies and decisions made. There are three important conditions a rational planning exercise should fulfil. First, the reasons for making planning decisions should be carefully thought through – decisions should be arrived at by considered reflection rather than by guesswork, 'hunch' or intuition alone. This in turn implies, secondly, that the reasons for making planning decisions should be explicit. If someone were to ask a rational planner why a certain decision had been made, the planner should be able to explain the reasons which led to that decision. Thirdly, if the whole *process* of planning is rational then *each and every stage* of a planning process should be carefully and explicitly thought through. The problems or objectives a plan is trying to solve or achieve (i.e. the first stage of the planning process in Figure 4.2) should be carefully considered and explicitly stated, and likewise the alternative strategies and so on with all the other stages.

The rational process model as a normative model of planning

So far only a *description* of the rational planning model has been given; from this it does not necessarily follow that we *ought* to make decisions in this carefully considered and explicit way. Thus some people hold that some kinds of decisions are best made on the basis of intuition – for example, the decisions to marry someone, or have children, or follow a particular career – are all clearly very significant decisions yet some might argue that these are best made on the basis of intuition and 'feeling'. However, where decisions are being made which significantly affect the lives of others and where there is some form of collective action to implement such decisions, there are strong reasons for insisting that decision-making and action should involve both careful and explicit deliberation before policies and plans are adopted (e.g. on the grounds of the numbers of people likely to be affected, the fact that these effects may last a lifetime, etc.). Because town planning is a form of social action, the model of rational planning represents more than just a description of what it might be like to plan rationally; it also represents a model of how we *should*, ideally, go about planning. The rational process model, therefore, suggests itself as a normative model or theory of planning and this was assumed by the rational process theorists of the 1960s and 1970s. As Faludi (1973a, p. 116) put it: 'It is only as a *normative* model that the rational planning process has any meaning at all.'

The rational process model as a model of instrumental (means–end) reasoning

As noted above, the rational process view of planning is clearly a process or procedural theory of planning, not a view or theory about the object or 'substance' of planning (in Faludi's terms, it is a procedural and not a substantive planning theory). Hence the rational process model describes only the 'form' (the 'procedure') of the reasoning involved in making rational decisions; it says nothing about the actual 'substantive' ends or goals planning should aim at. This model simply tells us that, to be rational, a process of decision-making must identify *some* problem or objective to be solved or aimed at, *some* plan designed to solve the problems or achieve the objectives and so on. It tells us nothing in itself of *which* problems to solve, *which* plans to make, etc. This will be determined by the particular situation. The rational process view of planning is therefore about the method or 'means' of planning, not about the 'ends' of town planning. This is a 'means–end' or 'instrumental' model of reasoning, not a model of substantive moral reasoning. Because it was 'merely formal', and said nothing about what ends planning should seek to achieve, it was later to come under criticism (see Chapter 6).

The debate about 'disjointed incremental' versus 'rational comprehensive' planning

Finally, there has been debate about whether the rational process view of planning necessarily implies the adoption of a 'comprehensive' approach to planning and policy-making. At first sight, it might seem that rationality does require comprehensiveness, for in order to make perfectly rational decisions, one needs to consider *all* possible alternatives. However, limited time in which a decision has to be made, a lack of resources to examine all possibilities, etc., mean that, in reality, it is often impossible to be thoroughly comprehensive. Decision-makers and planners may also simply lack the capacity to absorb and make sense of all the relevant information; they may end up becoming more confused and thus less rational – problems raised by Charles Lindblom (Lindblom, 1959; Braybrooke and Lindblom, 1963). Lindblom proposed an approach which, he claimed, was more relevant to the real world of planning and policy-making. He suggested that, in most situations, planning has to be piecemeal, incremental, opportunistic and pragmatic, and that planners who did not or could not operate in these ways were generally ineffective. In short, Lindblom presented a model of the 'real world' planning as necessarily 'disjointed' and 'incremental', not 'rational' and 'comprehensive'.

It might be thought that, as Faludi (1973b, p. 153) put it, 'disjointed incrementalism is highly descriptive of real-life planning', but this does not necessarily mean that this is how planning *ought* to be approached. Lindblom's critique, however, is not so easily disposed of. If real-life planning *can* only be disjointed and incremental, and since 'ought implies can' (i.e. a view about how we *ought* to act is only persuasive if we *can* act in this way), Lindblom's suggestion was that a disjointed incremental approach to planning was the *only possible* approach to planning. It was the best approach we could hope for – much better, in fact, than the impossible ideal of rational comprehensiveness.

The 'rational comprehensive' versus 'disjointed incremental' debate about planning is considered in Chapter 7. Suffice it to say here that the strength of Lindblom's critique rests in our acceptance that being rational requires being comprehensive. If a stranger entered my room with a gun in his hand then I would decide to act quickly. I would therefore have to assess quickly the alternatives open to me and then act. In so doing I would not be appraising my options 'comprehensively'. And yet, given my knowledge of what madmen with guns can do, my decision to forego comprehensiveness would be perfectly rational behaviour. However, we often cannot afford the luxury of examining every possibility because we haven't the time or resources, or because we can only absorb limited information. On such occasions it is perfectly rational to examine, and act upon, the few alternatives we initially perceive as best we can. Adherents to the rational model would say that 'as best we can' means going through something like the process of decision-making described earlier, however imperfectly. From this point of view, rationality does not require comprehensiveness: in certain circumstances it can be rational to 'go through' a rational process of decision-making quickly, even 'disjointedly' and 'incrementally'.

Faludi's (1973b, Chap. 8) suggestion that rational comprehensive and disjointed incremental planning are *alternative* modes of planning therefore poses a false dichotomy. Faludi himself gives various strategies for making rational decisions in conditions when it is impossible to acquire or absorb all relevant information (*ibid.*, Chap. 6). One was Amitai Etzioni's (1967) idea of 'mixed scanning', which distinguishes more fundamental or 'strategic' decisions from more detailed decisions, and then advocates concentrating the process of rational decision-making on the more fundamental decisions. The mixed-scanning approach also involves 'tracking' the detailed consequences of crucial, strategic decisions. In this way the capacity to oscillate – or 'scan' – between more general or strategic and more detailed levels of thinking is developed.

CONCLUSION: RATIONALITY, CYBERNETICS AND MODERNITY

This chapter has described two different theories of planning which came to the fore in the 1960s. If we accept Faludi's (1973, Chap. 1) distinction between *substantive* and *procedural* planning theories, then the systems view of planning, being a theory about the 'substance' (the environment) which town planning deals with, was a substantive theory, whilst the rational process view was clearly a procedural theory of planning.

That these two theories are distinct is shown by the fact that it would be possible to subscribe to one and not the other. Even if one had not heard of the systems view of planning, one might still believe that planning should be approached in a rational way, and so adopt the rational process model of planning described in this chapter. Indeed, one could imagine this rational process view being combined with the post-war physical design view of planning described in Chapter 1. The fact that the systems and rational process views emerged simultaneously, and were often put forward jointly in the new planning textbooks of the 1960s, was thus something of a coincidence, but not entirely so. For although they are conceptually distinct, these two theories shared some common assumptions which were prevalent in the 1960s.

In this chapter I described some of the general contextual factors which help explain why both these theories were taken up by planning theorists in the 1960s. Perhaps the most important of these was the general interest in science and its application to planning and policy-making generally. One way of describing the change in planning thought which occurred from the 1950s to the 1960s is that, whereas in the 1950s and before, town planning was seen as primarily an art, by the end of the 1960s it had come to be seen as primarily a science. Both the systems and the rational process views fed off each other in bringing about this change in planning thought. But this interest in science and its practical application had deeper roots.

The two theories of planning described in this chapter can also be viewed as sharing certain fundamental assumptions about the nature of the world and the possibilities for human progress within it. To begin with, both theories were based on a belief in the benefits of planning, and in this they assumed what Giddens (1994, p. 58) has termed a 'cybernetic' model of control.

These presuppositions were part of a more general set of assumptions which have come to be associated with 'modernism' – not just the modern movement in architecture and the arts, but rather a certain way of thinking about the world and social action which had developed in the European Enlightenment of the 18th century. Central to this was a belief in people's capacity to improve the quality of human life on the basis of a scientific understanding of the world. Throughout most of human history, people's lives had been governed by forces which seemed beyond human control and which could strike anyone at any time. The growth of scientific understanding opened up the prospect of humanity gaining greater control over the forces of nature and using them to human advantage – in medicine, in agriculture, or in fashioning the environment (it is no accident that landscape design flourished in 18th century Europe).

With this there naturally arose a greater optimism about the future – a belief that human life need not be inevitably subject to the whims of 'fate', but could be improved through rational understanding and action. Twentieth century modernist Utopias, such as Le Corbusier's 'radiant city', were expressions of this optimism, as was the more commonplace belief in 'progess'. So too was the belief that, with a proper scientific understanding of the environment as a 'system', coupled with the application of a rational method of decision-making and action, cities and the environment generally could be planned to improve – even 'maximise' – human well-being. In this, both the systems and the rational process views of planning were part of the heady 'modernist' optimism of the 1960s. Indeed, the systems and rational process views of planning can be regarded as marking the high tide of modernist thought – the crest of that wave of optimism about the use of science and reason for human progress which had formed the European Enlightenment of the 18th century.

NOTES

1. In spite of the logical distinction drawn here between these theories, they have sometimes been described jointly as if they were one and the same theory, and this has generated some confusion (e.g. Peter Hall may confuse the two theories in this way in Hall, 1975, Chap. 10). Confusion may also have arisen because the two theories were advanced simultaneously in the 1960s, and sometimes systems theorists, in particular, did not clearly distinguish them. Thus both McLoughlin (1969, Chap. 5) and Chadwick (1971, Chap. 4) incorporate accounts of planning as a rational process of decision-making while advancing the systems view of planning, as if the rational process view was part of the systems view (or *vice versa*).
2. The point here is analogous to that made by the philosopher of science Karl Popper, in relation to carrying out observations in scientific reasearch. As Popper insists (1963, Chap. 1), scientific research does not begin with empirical observation because any empirical observation is necessarily guided by some prior research problem, hypothesis or belief.
3. In other words, a central part of identifying and defining the problems (or aims) planning should address should involve a *critical* analysis of whether something assumed to be a problem actually is a problem.
4. Such systematic evaluation might also include some estimate of the probability of certain effects occurring, for some effects will be more uncertain than others. It should also include an analysis of *who* will experience the effects of different alternatives and to what degree for, again, different groups may stand to gain or lose from different alternatives.

5

Planning as a political process

MODERNISM AND URBAN PROTEST

Alongside the 1960s belief in progress based on reason and science sat another distinguishing mark of modernism, namely, an inclination to believe that the world could be made better by casting aside tradition and constructing things anew from 'first principles' based on 'pure' reason.[1] This anti-traditionalist tendency again had its roots in the European Enlightenment of the eighteenth century, and was given political expression by the American and French revolutions which sought to construct a new social and political order based on axiomatic principles (such as the rights of man) derived from reason rather than tradition.[2]

At the beginning of the twentieth century, the modernist desire to break away radically from the past was vividly expressed in European and American culture, in a startling outburst of new forms of artistic expression.[3] Even in philosophy, modern philosophers exhibited an inclination to reject the past and reconstruct the discipline on new premisses.[4] And this modernist tendency to reconstruct things anew was a central dynamic in the modern movement in architecture and planning.

This break with tradition was expressed in two ways. First, in its physical form, modern architecture was stripped down to bare, structural essentials, thereby rejecting the use of ornament or decoration which had been a central component of all past traditions of architecture, whether romantic or classical (see Figure 5.1). In modern town planning, the wish to create a neatly ordered urban form was the complement to the geometrical simplicity of modern architecture (See Chapter 2).

Secondly, modern town planning was characterised by 'Utopian comprehensiveness'; that is, a drive to build or rebuild anew whole cities or large parts of them. Utopianism always had this tendency, but what was new was the capacity of modern industrial technology to build some of these Utopian schemes on a large scale. Thus, within fifteen years of the end of the Second World War, a whole ring of new towns had been built around London and in the inner areas of many cities, huge schemes of comprehensive redevelopment had transformed the old urban fabric.

This drastic 'clean-sweep' redevelopment was most strongly felt in British cities in the 1960s and caused widespread public protest. Protest was aroused by two kinds of modern development in particular. First, there were the massive schemes of housing redevelopment, in which great swathes of 19th century

Figure 5.1 The stripped-down form of modern architecture – Adolf Loos's Steiner House in Vienna, 1910

terraces were swept away and replaced by a completely new environment of high-rise blocks of flats. Whilst some of the people subject to such schemes were happy to be rehoused from their cramped Victorian terraces to brand new flats, others objected to the dislocation this created (as Jon Gower Davies (1972) and Norman Dennis (1972) documented in relation to Newcastle upon Tyne and Sunderland, respectively).

The second form of protest was directed at the major schemes for new urban motorways to accommodate the rising tide of motor traffic. Marshall Berman (1982, p. 329) graphically summed up the 1960s as a struggle between 'the expressway world' and 'a shout in the street'.[5] Once again, the expressway world owed much to Le Corbusier's vision of the modern city. This vision was realised as concrete motorways sliced through people's homes and neighbourhoods. Whole communities were disrupted, either by being displaced altogether, or by ending up living next to roaring traffic. Thus residents of a terrace in Bristol, who once enjoyed a view across a green park, found they were looking out of their bedroom windows directly into the elevated section of the M32 as it carved its way into the centre of Bristol. In this urban nightmare, people's protests sometimes took the form of a desperate cry for help. In London, where the new elevated 'Westway' was built within feet of people's housing, residents hung out a banner from their bedroom windows saying simply: 'GET US OUT OF THIS HELL'.

These urban protests highlighted something which the systems and rational process theorists had downplayed or overlooked altogether: plans and planning decisions rest upon value judgements about what kind of environment it is desirable to create. The force of the protests showed that such judgements were often not (as an earlier generation of planning theorists like Keeble had assumed) just 'common sense'. They were matters of intense, sometimes passionate, contention and disagreement. In other words, town planning judgements were not so much technical or scientific as political. It was even suggested that it was misleading to describe town planning as a 'science'; instead, it would be better described as a form of political action directed at realising certain values. Clearly, then, this idea was somewhat different from the view of planning promulgated by the systems and rational process theorists of the 1960s.

In this chapter I describe the emergence and evolution through the 1960s of this view that planning is inherently a normative and political activity. I discuss how both the systems and the rational process views of planning tended to overlook the value-laden and thus political nature of town planning, and how they contributed to a 'technicalist' view of planning which concealed its political content. I examine some of the theoretical literature of the 1960s which was alive to the political nature of town planning, and I conclude by discussing the idea of 'public participation' in planning. Here, I examine the government response in Britain to the political contentiousness of town planning, which found expression in the 'Skeffington Report' on public participation in planning, as well as a more radical view of participation.

SCIENCE, VALUES AND RATIONALITY: FALSE TECHNICISM

The suggestion that town planning was 'political' was not completely new, for it had been recognised before that planning embodied political judgements and decisions – the very establishment of a town and country planning system involved government intervention and legislation, and in 1947 (as at other times) this had excited some political controversy, especially over the extent to which there should be public control over private property rights. Moreover, under the 1947 system, both development plans and decisions controlling development had to go through the normal channels of representative democracy and be approved by local councils or, in the case of development plans, central government.

However, as noted before, post-war planning theorists like Keeble tended to assume that the 'principles' of planning were self-evident and commanded general assent. The social democratic 'mixed economy' – the 'middle way' between capitalism and socialism – seemed so obviously sound, that as already mentioned, some political theorists, like Daniel Bell (1960), spoke of the 'end of ideology'. It was perhaps understandable, then, that issues of public concern, such as the state of cities, were seen in 'technical' terms – as 'problems' that could be 'solved' with appropriate scientific understanding and the application of rationality. However, although the systems view implied that the environment (e.g. cities) should be studied and analysed scientifically, as 'a system' of interconnected activities, one could adopt this perspective and still acknowledge that it was a matter of moral and political debate what the 'optimal' state of any given system might be. And the rational process view, seemed to make explicit the value-laden nature of planning by specifying the goals and objectives of any planning exercise, by acknowledging that there could be alternative plans and policies and, if there were, by stressing the need to evaluate these alternatives.

However in the 1960s both the systems and rational process views of planning came to be described as specialist 'technical' approaches to planning, as if their adoption could lead to the identification and solution of urban problems independent of considerations of values or political debate. This apolitical 'technicism' was evident in the way, for example, systems theorists wrote of 'modelling the urban system' and intervening to 'optimise its functioning', as if the city were some kind of a machine that had one politically uncontentious optimal state.

Moreover, even when planning theorists acknowledged the need to specify the goals of any planning exercise, and thereby acknowledged that planning was based on values, they were still inclined to speak as if the identification of the goals of planning was an expert, technical matter rather than a matter of *debate* about values and politics. This is seen in George Chadwick's (1971) account of goal formulation. Whilst acknowledging that goal formulation was 'the very crux and hinge-pin of the rational planning process' (*ibid*., p. 120), Chadwick described 'goal formulation' as if it were an exercise in which planners could be technically more expert than members of the public or their elected representatives:

> The clients of planners have never given the professionals in their employ any but the vaguest kind of goals . . . This throws a considerable responsibility upon the planner: he largely has to determine the goals of planning because his clients do not give them to him . . . We repeat our earlier suggestion: for planning to replace, or to add to, the normal processes of social choice in a democracy . . . one of the most forceful arguments for placing primary responsibility for goal formulation on the planner . . . [is] . . . the assumption, traditional to professionals, that, in some way, they 'know more' about the situations on which they advise than do their clients. (*Ibid*., pp. 120–1)

In the context of a political democracy this is an astounding statement, more reminscent of something one would expect to have been written by a state planner in the former USSR. Yet it captures, albeit in an extreme form, how even within western liberal democracies planning came to be seen as somehow above politics – as something that might, as Chadwick candidly admits, 'replace' politics. There really was something in Daniel Bell's claim after all, that this was an era in which ideology seemed to have been superseded.

A 'technicalist' interpretation of the rational planning process was also exhibited in relation to plan evaluation, and this is worth a more extended discussion.

Plan evaluation

'Plan evaluation' refers particularly to that stage in a planning process where alternative policies or plans are compared to decide which is preferable. The rational process and systems theorists of the 1960s tended to see the task of evaluation in terms of certain *techniques* for evaluating alternative plans on the assumption that such techniques, if properly operated, could provide the 'right answer'.

The commonest technique employed was cost-benefit analysis or some variant of it, such as Lichfield's planning balance sheet.[6] Both approaches try to itemise, and weight, the various costs and benefits that are likely to flow from the implementation of a plan. These are then entered on to a balance sheet with all the costs listed in one column and the benefits in the other. Each column is summed to provide an overall total of costs and benefits. This is done for each alternative and, in general, the preferred one was assumed to be that which yielded the greatest ratio of benefits to costs – in other words, that plan which 'maximised' benefits to the community affected by the plan.

The extensive literature on cost-benefit analysis (especially in welfare economics) includes sophisticated analyses of the methodological problems of operationalising cost-benefit evaluations: how individual costs and benefits may be identified and measured accurately (including allowances for 'discounting' over time); how to weight items reasonably which are not readily quantifiable in monetary terms, such as the beauty of a landscape or the loss of a habitat; what sense it makes to aggregate, or sum, entirely different kinds of costs and benefits; and so on. Lichfield's planning balance sheet was specifically developed and adapted for urban and regional planning to make more explicit the 'numerous important non-quantifiable and incommensurable items' relevant to plan evaluation in this field (Lichfield, Kettle and Whitbread, 1975, pp. 60–1).

Unquestionably, the task of having to specify and weight the effects – both harmful and beneficial – of any plan is essential to any attempt to evaluate alternative plans or policies. One cannot effectively compare the merits of different proposals without first having some idea of what the effects of each proposal will be. However, the idea that something like cost-benefit analysis (or Lichfield's variant on it) can provide an uncontroversially right 'answer' is seriously misleading, for the problem of deciding which alternative is *preferable* remains. This problem is not 'solved' by means of a statistical calculation, for the act of choosing the preferable option still remains a matter of *value*.

Yet to many planners it did (and still does) seem ethically uncontroversial that the overall purpose of planning was somehow to enhance and thus 'maximise' human welfare, and so it seemed obvious that the 'best' plan from a number of alternatives must be the one which produced the most benefits in relation to costs. Indeed, to most planners this seemed such obvious common sense that there seemed little point in dwelling further on this question of value, and so the main problems of plan evaluation naturally came to be seen to be the technical ones of how to identify the numerous costs and benefits, how to weight them appropriately, and so on. To the extent that town planning theorists thought this sufficient, they exhibited ignorance of value theory. For the idea that the best thing to do is that which *maximises* benefits (or welfare, happiness, etc.) for a given population as a whole derives, from a specific moral theory or philosophy – utilitarianism, which has been the subject of intense debate and controversy.

Utilitarianism was first expounded systematically by the eighteenth-century English philosopher Jeremy Bentham, who held that actions are morally justified when they bring about the greatest happiness (or 'utility') for the greatest number of people; that is, when they maximise happiness. Bentham devised a method of itemising, weighting and summing all the 'pleasures and pains' which would likely ensue from a course of action, and this he termed his 'felicific calculus'. Given the premise of maximising utility (or happiness, welfare, benefits, etc.), Bentham believed that his calculus of pleasures and pains ('costs and benefits' in the modern jargon) could provide answers to moral problems. Modern cost-benefit analysis and Lichfield's planning balance sheet are simply up-to-date versions of Bentham's original calculus. As Lincoln Allison (1975, p. 74) observed: 'The principles which for the most part dominate modern planning . . . in Britain are to be found in a single book: Jeremy Bentham's *Principles of Morals and Legislation*'.

Commonsensical though Bentham's utilitarianism may seem to be, it is important to appreciate that it is just one moral theory amongst others, and that whilst there have been many philosophers who have continued to subscribe to and further refine utilitarianism, there have been many others who have been critical. A standard criticism of utilitarianism is that the principle of maximising welfare over a population as a whole does not consider how the costs and benefits arising from an action should be *distributed* amongst a population. Thus it is possible to imagine a planning policy (e.g. for a new road) which would maximise the welfare of a majority of a city's population, and yet distribute the costs and benefits arising from this development so that sections of the population who are already disadvantaged bear most of the costs, whilst those who are already better off enjoy most of the benefits. Many of the urban motorways built in the 1960s had precisely this inegalitarian effect, for they were often routed through poor areas where the local inhabitants ended up bearing the brunt of the costs (the disruption arising from development, the noise and pollution, etc.) even though few of the inhabitants of these areas owned cars to benefit from using the new road.[7] If a plan or policy were to result in such a distribution of costs and benefits then it would strike many as unjust, even if it achieved a net benefit for a city's population taken 'as a whole'. Many might hold that a proposal which distributes costs and benefits more equally, or which even imposes greater burdens on the better-off, would be ethically preferable, even if it scores less than a proposal which 'maximises' a population's welfare. Other principles of distributive justice have therefore been advanced, and the most widely discussed of these in recent times has been the theory of justice advanced by the American moral philosopher, John Rawls (1972).

This is not the place to pursue the philosophical debate over the relative merits of different moral principles to guide action. But we have shown that, even when the costs and benefits which are likely to flow from various alternatives have been itemised, this itself does not necessarily resolve the question of which alternative is preferable. That judgement still requires an overall value judgement about what it is best (right, just, etc.) to do, and this cannot simply be dealt with by means of a 'technique', be it cost-benefit analysis or any other.

Against this background, it is significant that the systems and rational process theorists of the 1960s showed little awareness that the methods of plan evaluation they recommended were based on particular, and debatable, ethical positions and principles. In the 1960s it was not acknowledged, for example, how cost-benefit analysis rested on Benthamite utilitarianism or even that the preferred methods of evaluation were in any way 'utilitarian'. In an entire book devoted to 'evaluation in the planning process', Lichfield and others do not even mention utilitarianism even though the moral principle of utility underpins all the methods of evaluation they discuss (Lichfield, Kettle and Whitbread, 1975).[8] It is therefore not surprising that cost-benefit analysis, and other methods of plan evaluation, came to be seen and conveyed to planners as *techniques* for evaluating alternative plans rather than, as they should be, *sources of information* for, and thus *aids* to, plan evaluation.

Cost-benefit analysis is only one illustration of a more general and deeper mistake made by some planning theorists in the 1960s. There was a general tendency to conceive of policy-making and planning as 'sciences' which employed 'scientific' and thus value-free techniques. The very term 'policy science' betrayed this. As Laurence Tribe (1972, p. 75) pointed out:

> One of the most persistent beliefs about the techniques [of policy science] . . . is a conviction of their transparency to considerations of value and their neutrality with respect to fundamental world views . . . the myth endures that the techniques *in themselves* lack substantive content, that intrinsically they provide nothing beyond value-free devices for organising thought in rational ways – methods for sorting out issues and objectively clarifying the empirical relationships among alternative actions and their likely consequences. The user of such techniques, the myth continues, may turn them to whatever ends he seeks. Ends and values, goals and ideologies are seen as mere 'inputs' to a machinelike, and hence inherently unbiased, process of solving problems consistent with the facts known.

As Tribe demonstrates, even the very *method* of cost-benefit analysis involves assumptions about the nature of problems (or 'costs') which embody a particular value stance or ideology. For example, the very disaggregation of costs and benefits into individual 'bits' necessarily oversimplifies and distorts the very nature of some kinds of costs and benefits. Cost-benefit analysis does not, therefore, even provide a scientifically objective characterisation of some of the policy effects which it is the whole purpose of this analysis to identify and weigh.

Planning, Popper and scientific method

As we saw in the previous chapter, the reconceptualisation of town planning in terms of systems theory brought with it the baggage of 'modelling', quantification ('mathematical' modelling) and the use of computers to model complex systems. These seemed to be the hallmarks of 'being scientific', and there was also much talk of the scientific *method* in relation to planning. Andreas Faludi, the leading theorist of rational planning in Britain in the early 1970s, even claimed that 'Planning is the application of scientific method . . . to policy-making' (Faludi, 1973a, p. 1; see also Jay, 1967).

Thus town planning students in the 1960s and 1970s were advised to read Karl Popper's pioneering work (Popper, 1957, 1963) on scientific method. The rational process of planning was sometimes likened to Popper's account of the scientific method, with planning goals or policies seen as analogous to scientific hypotheses which should be subjected to rigorous testing before adoption. As Bryan Magee (1973, p. 75) put it when writing about Popper in relation to policy: 'A policy is a hypothesis which has to be tested against reality and corrected in the light of experience.'

This analogy between the scientific method and the rational planning process was (and remains) in some ways fruitful. There are two important aspects of Popper's account of science. First, Popper insisted that the pursuit of scientific knowledge did not begin with empirical observation from which theories were then inferred. Rather, it is the other way round. All empirical investigation in science is driven by some prior idea or belief about some aspect of the world, even if this is just a vague 'hunch' or guess. Secondly, empirical observation tests the truth content of a given hypothesis or theory. For Popper, however, the proper method of testing scientific hypotheses is to try to find empirical evidence which *falsifies* them. If such evidence comes to light then clearly we need to reject or revise our initial hypotheses. Scientific knowledge, therefore, advances by identifying and correcting *mistakes* in one's beliefs about the world; it advances by trial and error, by conjectures and *refutations* (Popper, 1963).

Popper's emphasis on the need to articulate the initial conjectures or hypotheses which guide scientific research is analogous to the need, in planning, to specify at the outset the problems or goals one is seeking to solve or achieve, for if one isn't clear about these to begin with then a planning process has no rationale; it is, literally, aimless. Similarly, Popper's stress on the need to subject one's initial conjectures to rigorous 'falsifying' tests suggests that planning problems, goals or policies should likewise be subjected to rigorous testing. Had some of the practices of modern town planning, such as mass high-rise housing, been subjected to such 'Popperian' testing before being put into practice, they might never have been adopted and much misery would have been avoided.

Though fruitful, the analogy between planning and science needs to be treated with some caution, however. Science is concerned with describing and explaining aspects of the world as a matter of *fact*. In this respect it is *descriptive*. Town planning, on the other hand, seeks to be *prescriptive* because it is concerned with intervening to change some aspect of the world to improve it. Planning is therefore engaged in altering given facts. Scientific investigation describes and explains the phenomena affecting the environment but town planning is concerned with trying to alter these facts, for example, by putting forward proposals to regenerate decayed inner city areas, making transport plans which enable people to travel around cities with ease whilst simultaneously reducing pollution, etc.

> the whole point of personal or social choice in many situations is not to implement a *given* system of values in the light of the perceived facts, but rather to define, and *sometimes deliberately reshape*, the values – and hence the identity – of the individual or community that is engaged in the process of choosing. (Tribe, 1972, p. 99, emphasis added)

It follows from this that the prime questions facing any planning exercise are questions about what it is best to do, and these are questions of *value*. Town planning is not a science in the usual sense of that word. It is therefore more accurate to define planning as an evaluative or *normative* activity. Moreover, since town planning action can significantly affect the lives of large numbers of people, and since different individuals and groups may hold different views about how the environment should be planned, based on different values and interests, it is therefore also a *political* activity. The planning theorists in the 1960s who saw planning as a science therefore misconceived the very activity they were seeking to describe.

EARLY THEORISTS OF PLANNING AS A NORMATIVE POLITICAL PRACTICE

Some planning theorists were alert to the value-laden and political nature of planning and therefore saw that planning should not be construed purely as a technical or scientific activity. As noted before, two British writers, Jon Gower Davies (1972) and Norman Dennis (1972), drew attention to the political nature of planning in their work on comprehensive housing redevelopment. However, it was mostly American planning theorists who first articulated its political nature. One of the first was Norton Long (1959, p. 168):

> Plans are policies and policies, in a democracy at any rate, spell politics. The question is not whether planning will reflect politics but whose politics it will reflect. What values and whose values will planners seek to implement? . . . plans are in reality political programs. In the broadest sense they represent political philosophies, ways of implementing differing conceptions of the good life. No longer can the planner take refuge in the neutrality of the objectivity of the personally uninvolved scientist.

Two other Americans, Martin Meyerson and Edward Banfield, whose investigation of housing policy in Chicago (1955) revealed that different groups in that city had significantly different interests with respect to housing, and thus differing views about the ends which the city's housing policy should pursue, have also been mentioned. Both were advocates of a rational process of planning; indeed, Faludi (1973a, p. 115) credits Banfield with being the first theorist to introduce the rational decision-making process into the planning literature. However, in advocating the rational process model they were clear that the definition of the ends planning should pursue, together with the possible alternatives, was central to the planning process; and that these things were contested matters and thus matters of political debate and decision.

Two other Americans, in the early 1960s, Paul Davidoff and Thomas Reiner, emphasised the value-laden and hence political nature of planning. In a paper entitled 'A choice theory of planning' they also view planning as a 'process', but emphasise that throughout it is a process of *choice*.

> The choices which constitute the planning process are made at three levels: first, the selection of ends and criteria; second, the identification of a set of alternatives consistent with these general prescriptives, and the selection of a desired alternative; and third, guidance of action toward determined ends. Each of these choices requires the exercise of judgement; judgement permeates planning. (Davidoff and Reiner, 1962, pp. 11–12, cited in Faludi, 1973a)

They stress that the first of these 'levels' of the planning process – the definition of the ends or aims of planning – necessarily rests on value judgements about what state of affairs it is desirable to plan for. And they point up the contentious, and hence political, nature of these value judgements by urging that a planner

> cannot, as an agent of his clients, impose his own ideas of what is right or wrong . . . Our contention rests on the thesis that goals are value statements, that value statements are not objectively verifiable, and, therefore, that the planner, by himself, cannot reasonably accept or reject goals for the public. This is crucial: we maintain that neither the planner's technical competence nor his wisdom entitles him to ascribe or dictate values to his immediate or ultimate clients. This view is in keeping with the democratic prescriptive that public decision-making and action should reflect the will of the client; a concept which rejects the notion that planners or other technicians are endowed with the ability to divine either the client's will or the public will. (*ibid.*, p. 22)

The contrast here with Chadwick's view quoted earlier is striking, and all the more so given that Davidoff and Reiner were writing nearly ten years before Chadwick. Furthermore, Davidoff and Reiner were not only alert to the way the value-laden nature of planning implied that planning was a political process; they also saw how this raised questions about the role and responsibilities of planners *vis-à-vis* the clients they served – questions, in other words, about the nature of 'professionalism' in planning.

In the above quotation Davidoff and Reiner imply that, in his or her work as a professional, the planner should confine him or herself to 'technical' matters concerning planning, such as making clear what the choices are, setting out the likely effects of adopting these alternatives and so on. The planner may get 'involved with values' (*ibid.*), but what values planning aims to realise should remain a matter for political, democratic choice.

This view of the planner's professional role implies a clear distinction between technical or factual matters on the one hand, and matters of value and political choice on the other. This is a distinction Davidoff and Reiner endorsed. As they put it (*ibid.*, p. 19): 'Our analysis of the value-formulation process and of the planner's resposibilities in dealing with values has as its basis the philosophical distinction between fact and value.' In spite of this, later in the 1960s Davidoff came to question this 'technicalist' view of the planner's role and to argue that planners should involve themselves more actively in the political process by acting as 'advocates' for client groups within the public, especially disadvantaged or minority groups whose interests were not well represented in the process of planning:

> The prospect for future planning is that of a practice which openly invites political and social values to be examined and debated. Acceptance of this position means rejection of prescriptions for planning which would have the planner act solely as technician . . . the planner should do more than explicate the values underlying his prescriptions for courses of action; he should affirm them; he should be an advocate for what he deems proper . . . The recommendation that city planners represent and plead the plans of many interest groups is founded upon the need to establish an effective urban democracy, one in which citizens may be able to play an active role in the process of deciding public policy. Appropriate policy in a democracy is determined through a process of political debate. The right course of action is always a matter of choice, never of fact. In a bureaucratic age great care must be taken that choices remain in the area of public view and participation'. (*ibid.*, pp. 277–9, cited in Faludi, 1973a)

We can see here that Davidoff had moved to a much stronger, or more committed, 'political' view of the planner's role, although it is interesting to observe that there remains some equivocation in his position. For whilst Davidoff suggests that the planner should seek to represent and plead for the plans of the 'many interest groups' in the public, and thereby facilitate the active involvement of citizens in 'deciding public policy', he also speaks of the planner explicating and affirming the values underlying 'his' prescriptions for action, and being an advocate for what 'he' deems proper.

The journey taken by Davidoff neatly illustrates the transition town planning underwent during this decade. At the beginning of the 1960s and before, most planners took little interest in politics, although in countries such as Britain where a system of statutory planning had been established, it was acknowledged that plans and planning decisions had to 'go through' the political process. This process was something of a formality because, so often, it was assumed that the planner, as a 'technical' adviser, 'knew best'.[9] By the end of the decade all the talk was of how 'political' planning was. And once this idea took hold, it seemed logical that the public should have the opportunity to 'participate' in planning, as Davidoff suggests.

PUBLIC PARTICIPATION

The recognition that planning decisions were 'political' naturally implied, in any political system purporting to be democratic, that the public should have some say in, or should 'participate' in, those decisions. This section examines how the idea of public participation in planning emerged and was dealt with in official (i.e. government) circles in Britain before we return to some more overtly theoretical work on the subject of participation.

The shortcomings of representative democracy

In one sense the idea and practice of the public 'participating' in planning was not new, for the planning system instituted by the Town and Country Planning Act 1947 had provided for members of the public to voice their views. Thus local planning authorities were required to publicise applications for planning permission and, in particular, to consult immediate neighbours of such proposals. They were also required to publicise the submission of development plans to the minister, and members of the public could inspect these plans and make any objections by writing to the minister. The public inquiry system also operated with respect to both the submission of development plans and appeals against planning decisions. On top of all this, there were the normal channels of representative democracy by which the public elected politicians to represent their interests in both central and local government. At local government level, all local authorities had planning committees whose job it was to oversee and make decisions about plans and planning applications, and these meetings were open to the public who, obviously, could make representations about any planning matter through their local councillor. So, as Michael Fagence (1977, p. 258) pointed out: 'Although the attitudes towards citizen participation throughout the 1940's and 1950's were essentially conservative, the legislative framework within which British planning was conducted incorporated at particular points in the process opportunities for some types of participation.'

However, the intensity with which some planning schemes were opposed by some groups in the 1960s suggested that the prevailing mechanisms of representative democracy were not working very sensitively. The idea that 'planning is political' thus came to possess a stronger connotation and led to calls for the public to have the opportunity to become more actively involved. It was in this context that the idea of 'public participation' in planning emerged. As Gyford (1976, p. 143) put it: 'Historically the election of representatives has been the primary device for enabling the public to participate in decision making in local government . . . During the 1960s Britain shared with other Western industrialised societies the emergence of demands for more effective mechanisms of participation than those traditionally employed'.

The government response to public participation

In Britain the idea that the public should 'participate' in planning was first raised in 1965 by the report of the Planning Advisory Group (the 'PAG' report) which had been set up by the government. At the beginning of the report the PAG set out four objectives, the first of which was 'to ensure that the planning system serves its purpose satisfactorily both as an instrument of planning policy and *as a means of public participation in the planning process*' (Ministry of Housing and Local Government, 1965, p. 2, emphasis added).

As we saw before (Chapter 3), the report was mainly concerned with examining the system of development plans which had been in operation under the Town and Country Planning Act 1947. The report's main recommendation (which was subsequently adopted in the Town and Country Planning Act 1968) was to distinguish between two levels of development planning – strategic or 'structure' planning and 'local' planning. However, in making this distinction the report also drew a distinction between two levels of responsibility for making the final decisions about plans. Thus the strategic structure plans, though prepared by local planning authorities, were to be approved by the minister responsible for planning at central government, whilst local plans would become entirely a 'local responsibility' (*ibid.*, para. 7.3, p. 44). In this way the PAG hoped that local-level development plans would 'make for better and more effective planning at the local level and a greater degree of public participation in the process' (*ibid.*, para. 7.4, p. 45). Indeed, the report suggested that local planning authorities, in preparing their local plans, 'must provide an opportunity for local comment or objections to be made and must consider these views before adopting a plan' (*ibid.*, para, 7.3, p. 44).

However, quite how the public might participate in planning in ways which differed from those traditionally employed remained unclear. In March 1968, therefore, while the 1968 Act was being prepared, the government minister responsible for planning established a special group under the chairmanship of A.M. Skeffington to 'consider and report on the best methods, including publicity, of securing the participation of the public at the formative stage in the making of development plans for their area' (Department of the Environment, 1969, p. 1). The report (the 'Skeffington report') was published in 1969.

As noted earlier, even in Davidoff's 'radical' political view of planning there was some equivocation between the view that the public should decide public policy and the view that planners themselves should prescribe what *they* thought right or appropriate. We see this equivocation in the Skeffington report too, where, on the very first page (*ibid.*, p. 1), the authors explain what they mean by 'participation':

> We understand participation to be the act of sharing in the formulation of policies and proposals. Clearly, the giving of information by the local planning authority and of an opportunity to comment on that information is a major part in the process of participation, but it is not the whole story. Participation involves doing as well as talking and there will be full participation only when the public are able to take an active part throughout the plan-making process. There are limitations to this concept. One is that resposibility for preparing a plan is, and must remain, that of the local planning authority. Another is that the completion of plans – the setting into statutory form of proposals and decisions – is a task demanding the highest standards of professional skill, and must be undertaken by the professional staff of the local planning authority.

Here we see Skeffington insisting, on the one hand, that public participation involves more than just being informed and responding to that information, more than just 'talking'; it involves taking an 'active part' in plan-making. On the other hand, Skeffington says there are 'limitations to this concept', and that the task of and responsibility for preparing plans 'must remain' with professionally trained officers (i.e. planners) and the local authority, respectively.

The report proposed some interesting ideas for fostering the more active involvement of citizens, such as 'community forums' to liase with local planning authorities and the idea of appointing 'community development officers' to reach out to those groups who tended not to get actively involved. Notwithstanding these suggestions, the overall picture of participation is of local planning authorities, advised by professional planners, retaining the ultimate responsibility, and thus power, to prepare and take decisions. In other words, participation is primarily seen as involving more *consultation* with the public rather than the public actively *participating* in decision-making.

Arnstein's ladder of participation

The notion that public participation meant 'consultation' was contested perhaps most famously by the American Sherry Arnstein in a much-quoted article which appeared in 1969. In this, Arnstein set out what she termed a 'ladder' of citizen participation (see Figure 5.2). Arnstein's ladder represents a lovely piece of 'conceptual analysis' of 'public participation'. What she shows is that public participation is not necessarily just one thing but rather it can be interpreted in several different ways; it can 'mean' different things. In particular, Arnstein drew attention to the crucial feature that, there are *'degrees'* of participation; there can be more or less participation (in this it is like democracy, of which participation is a sub-concept). The important question, therefore, is how much, or to what degree the public should be given a say in, and beyond that real power to decide, their affairs.

If we think of democracy and participation as ranging along a continuum as Arnstein suggests, then there is of course a position at one extreme of this continuum where the public effectively has no say or power at all (for the sake of completeness Arnstein indicates this by the bottom two rungs of her ladder). Moving up we come to forms of participation in which citizens are at least informed about what an authority is doing and, more than this, where citizens are consulted about an authority's ideas or proposals but where the authority itself retains the right to make the final decisions. Obviously, an authority in these circumstances may still decide to do things contrary to expressed public wishes (because ultimate power of decision resides with the authority). Arnstein thus terms rungs 3–5 on her ladder, somewhat cynically, degrees of 'tokenism'. The rungs above this (6–8) involve progessive transfers of power to citizens with, at the extreme of maximum participation, 'citizen control'.

If we set aside the non-democratic option of no participation, even for democrats the question of which level or degree of participation is most appropriate or desirable, either in relation to planning or other affairs, remains a

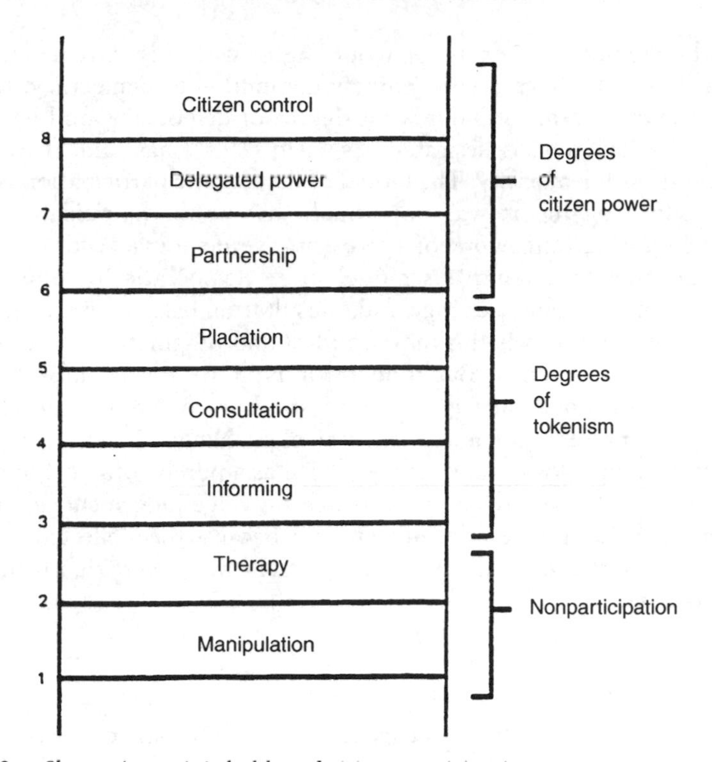

Figure 5.2 Sherry Arnstein's ladder of citizen participation
Source: Arnstein, 1969. Reprinted by permission of *The Journal of the American Institute of Planners*, Vol 35, July.

matter of debate. One could say that the question of 'how much' democracy (or participation) is either feasible or desirable has been the central debate in democratic theory (and arguably in political philosophy generally) since Plato's *Republic.*[10] The debate about public participation in planning is thus part of a more general philosophical debate about democracy. In the light of Arnstein's analysis, the conception of public participation as advanced by the Skeffington report was essentially one of consultation with the public. It did not conceive of or advocate participation as 'citizen control'.

In the 1960s, communities protested more vigorously against 'clean-sweep' planning, which was part of a wider political protest in favour of civil rights, against nuclear weapons and the Vietnam war, against universities, against capitalism. Some political and planning theorists argued for some form of participatory democracy, and Arnstein was one of these (see also Goodman, 1972). In her 1969 article she situates her analysis within the context of the student demonstrations in Paris in 1968, the civil rights marches by black people in the USA and campaigns by the 'have-nots' generally: 'There is a crucial difference between going through the empty ritual of participation and having the real power needed to affect the outcome of the process . . . the fundamental point [is] that participation without redistribution of power is an empty and frustrating process for the powerless' (*ibid.*, p. 216).

Not all planning theorists did or would agree with this extreme view. Even amongst those who have been genuinely committed to democracy, there has been intense disagreement about what degree of democracy, and hence what degree of participation, is desirable. As Held (1987) has said, there are different 'models' of democracy. The model of democratic participation advanced by the Skeffington report was one which saw public participation nesting within the existing framework of representative democracy; for Skeffington, participation meant making *this* model more democratic by improving the processes of information exchange and consultation between planners and the public. In the event, it was this more modest and pragmatic concept of participation which triumphed. But modest or not, the formal introduction of mechanisms for public participation into statutory planning – however successful or otherwise – was an expression of a changed concept of planning itself. Town planning was no longer regarded as a purely technical or scientific exercise. It was acknowledged that it rested on value judgements about desirable futures, and that these value judgements, because they reflected or affected the interests of different social groups in different ways, were rightly matters of political debate.

NOTES

1. As Giddens (1990) points out, the rejection of tradition has been a central feature of modernity.
2. This eighteenth-century political radicalism was associated with liberal thought, which in turn reflected the new economic order of capitalism. In the nineteenth and twentieth centuries, liberal capitalism was criticised by socialists. But the idea that society could be made better by 'revolutionary' change persisted as a central feature of socialist thought.
3. For example, cubist, fauvist, surrealist and abstract painting; unpunctuated 'stream of consciousness' writing and 'free verse'; atonal music; 'modern' dance; etc. A key driving force in all of these movements was to challenge, and indeed overturn, traditional ways (or subjects) of painting, writing, music making, etc.
4. Thus in Vienna in the 1920s, the logical positivist philosophers sought to reconstruct philosophy entirely on the basis of statements whose meaning could be rationally verified. According to them, there were only two classes of meaningful, verifiable statements: the propositions of logic and empirical science. All other kinds of propositions were dismissed as meaningless, and on this basis most traditional 'metaphysical' philosophy was rejected as nonsense (see Ayer, 1936).
5. Berman was writing about the 1960s in New York, but what he says of New York was also true of Britain.
6. See, e.g. McLoughlin, 1969, Chap. 10; Chadwick, 1971, Chap. 11; Lichfield, Kettle and Whitbread, 1975.
7. This was exactly the case with the M32 in Bristol mentioned earlier in this chapter which, together with a new inner ring road, mostly cut through working-class districts.
8. To be fair, the utilitarian background to Lichfield's 'social' cost-benefit analysis, or what he now calls 'community impact evaluation', is acknowledged in Lichfield's most recent book (see Lichfield, 1996, pp. 189–92).

9. For discussions of the relationship between unelected officers (such as planners) and elected members (i.e. local politicians or councillors) in British local government, see, e.g. Gyford, 1976, Chap. 2; Dunleavy, 1980, Chaps. 4 and 5.
10. For more recent philosophical discussions of this issue, see, e.g. Pateman, 1970; Walzer, 1970; Macpherson, 1977; Graham, 1986; Held, 1987.

PART III
PLANNING THEORY FROM THE 1970s TO THE 1990s

6

Theory about the effects of planning

INTRODUCTION

The theory of planning as a rational process of decision-making and action, coupled with the systems view of planning, continued to dominate planning theory into the 1970s. By the 1970s the rational process view of planning was seen as coming under the more general heading of 'procedural planning theory' (i.e. theory about the process of planning). As noted in Chapter 4, because procedural planning theory was, literally, theory of planning itself, Andreas Faludi (1973b) claimed that planning theory was synonymous with procedural planning theory. The title of Faludi's book – *Planning Theory* – had the same definitive finality about it as Keeble's *Principles and Practice of Town and Country Planning* twenty years before. There continued to be debates about rational planning – debates, for example, about 'rational comprehensive' verses 'disjointed incremental' planning, or about Etzioni's 'mixed scanning' as a 'third' approach to rational decision-making (see Chapter 4) – but these operated within a general acceptance of the view that planning theory was about planning seen as a process of decision-making.

That the process of planning necessarily involved political judgements could readily be incorporated into the rational process view. Thus it was now acknowledged that political judgements had to be made during the planning process and were most crucial at the point of defining what the goals of a planning exercise should be. And if the public should be given the opportunity to participate, then such participation could be 'built into' this or any other stage of a planning process. As Faludi (1973b, Part 4) put it, there could be a 'politics of rational planning', such that the political process could be incorporated into a system whose overall structure retained the rational steps of defining problems or goals, identifying and evaluating alternative plans and so on.

So settled was this view of planning in the early years of the 1970s that it almost seemed as if the problems of specifying what planning theory was about had been solved – that the theory of town planning had at last come of age. However, this comfortable view of planning theory was soon to be shaken by a wave of criticism. If we regard the criticisms levelled at town planning theory in the late 1950s and early 1960s (see Chapter 3) as the 'first wave' of criticisms of post-1945 planning theory, then the criticisms of procedural planning theory which emerged in the 1970s can be described as the 'second wave' of criticisms. Two main criticisms of this second wave can be distinguished.

First, procedural planning theory was criticised for its abstractness and generality – for offering 'merely' an extended definition of planning and not saying anything about how planning in practice operated or what its effects were, etc. It was thus criticised for lacking any 'content' or 'substance', for being 'empty' or 'vacuous' (see, e.g. Scott and Roweis, 1977; Camhis, 1979; Thomas, 1979). Secondly, the rational model of planning was criticised (e.g. Friedmann, 1969) for presenting a false 'top-down' view of planning which showed little understanding of the 'action end' of planning, including how plans and policies were or were not implemented.

The second of these criticisms which related to the implementation of planning and which came to preoccupy a growing number of planning theorists in the late 1970s and 1980s are discussed in Chapter 7. Central to the first criticisms was the view that planning theory should develop an understanding of the role and effects of existing planning. In the next section I discuss this first criticism, and in the rest of the chapter, describe research and theory in the 1980s, concerned with developing an empirical understanding of planning. As we shall see, the question of what effects were produced by institutionalised planning practice turned out to be a matter of theoretical controversy that dominated debates about planning theory in the second half of the 1970s.

CRITICISMS OF 'CONTENTLESSNESS' OR 'EMPTINESS'

Criticisms of 'contentlessness' or 'emptiness' of the rational process view of planning, were stated in unequivocal terms by Scott and Roweis (1977, p. 1098):

> Much of the existing literature on urban planning theory proceeds by adopting a 'theoretical' perspective that treats urban planning as an *abstract analytical concept* rather than as a sociohistorical phenomenon. Accordingly, conventional planning theory tends to proceed by initially positing formal or linguistic definitions of planning that have a purely *a priori* or nominalistic character. For example, one repeatedly encounters in the literature such global descriptions of planning as these: 'Planning is a goal-orientated process that seeks to achieve specified desired objectives subject to given constraints'; or, 'planning is an attempt to apply rational decision-making procedures to the guidance of social change'; . . . Now statements of this sort (that is, formal linguistic definitions) are in one sense unexceptionable. Yet this precise quality of unexceptionableness is gained at the cost of any sort of substantive specificity or predictive power. They may well be unassailable statements, but this is only by reason of their essential vacuousness . . . they tell us very little that is worthwhile about the empirical domain of urban planning'.

These points were echoed by Marios Camhis (1979). Commenting on a claim by Faludi that a key thing in planning is to establish a sound process of planning, Camhis (*ibid*., pp. 5–6) concludes that this concern leads the planner to turn away from the substantive issues it should be the purpose of planning to address:

> We are . . . led to believe that the right form of the planning process will inevitably determine the right content or what the real problem is. A policy on housing does not emerge out of an analysis of the situation, from any consideration of the needs of the people, and from the people themselves. The policy, according to Faludi, can only be successful if it conforms with the criteria that specify the 'right' planning process. Following this line of reasoning the planner is . . . nothing. Too much preoccupation with procedure or method in the abstract tends to push aside the real issues.

In the same vein, and in a specific critique of Faludi's work, Michael Thomas (1979, pp. 13–14, cited in Paris, 1982) claimed that 'procedural theory is essentially "contentless" in that it specifies thinking and acting procedures but does not investigate what is the content of these'. Because of this, 'Faludi cannot offer an explanation for the products of planning agencies because he has very little to say about what planning is doing, i.e. effecting particular changes in the environment' (*ibid*., p. 20). Thomas further suggested that procedural planning theory depoliticised planning in two ways. First: 'Procedural rationality can be used to advance any goal . . . [Thus] The emphasis which procedural theorists give leads to a neglect of the purpose of planning' (*ibid*., pp. 21–2; this reiterates Thomas's point that procedural planning theory is 'contentless'). Secondly, in advancing the model of rational decision-making as 'an ideal type . . . [as a] normative theory which seeks to improve the quality of planning' (*ibid*., pp. 15–16), procedural planning theorists were implicitly suggesting that rational planning should *replace* politics. So:

> [Faludi] asserts a positive value to be derived from a commitment to and involvement in rational planning . . . The political process will be substantially replaced by rational planning as the principal means through which people communicate with each other about the society they live in. Politics appears as an adjunct to planning. (*ibid*., p. 21)

In all these criticisms the same point recurs, namely, that procedural planning theory, and the rational process view of planning in particular, did not say anything about existing planning. It was merely 'abstract' or 'formal'; that is why it was contentless, empty, vacuous.

Implicit in this criticism was the suggestion that the theory of planning should be theory about planning in practice rather than about what rational decision-making might or should be like in ideal terms. These critics were calling for theory which described and explained what town planning does as a matter of fact. They were therefore calling for theory grounded in the *empirical* investigation of planning – in other words, the (social) *scientific* theory of planning. This was precisely the kind of understanding writers like Jacobs and Alexander had called for, and had criticised prevailing planning theories for lacking, over a decade before (see Chapter 3). Thus Jacobs's main criticism of planning theory at the beginning of the 1960s was that planning showed a lack of understanding of how the cities which planners were planning actually functioned. At the end of the 1970s, substantially the same criticism was still being voiced.[1]

In fact, the systems view of planning of the 1960s had developed partly in response to the earlier criticism of post-war planning theory for its lack of understanding of the places planners were planning (see Chapter 4). But in the late 1960s and early 1970s the systems view of planning (like the rational process view) was largely articulated in very general abstract terms. By the mid-1970s there were still few 'nose-to-the-ground' studies of how urban systems or parts of urban systems functioned.[2] Why was it, then, that ten years after Jacobs's and Alexander's call for closer investigations of actual cities, 'planning theory' continued to be expressed in these highly abstract 'conceptual' terms?

Part of the answer lies in the fact that the criticisms voiced at the beginning of the 1960s challenged a view of town planning which had held sway for a very long time. As noted in Chapter 1, the view of town planning as an exercise in physical planning and design had been the prevailing concept that stretched back hundreds of years. The criticisms which surfaced at the beginning of the 1960s thus required a fundamental rethink, or reconceptualisation, of what kind of an activity town planning was. It was therefore understandable that planning theorists went back to 'first principles' and that such fundamental theoretical work necessarily tended to be highly 'conceptual' and therefore 'abstract'. Jacobs and Alexander may have been right to castigate planners for their lack of understanding of real life cities, but implicit in that criticism was also the need for town planners to rethink completely what sort of an activity they were engaged in.

It is also worth adding that, given that the rational process view of planning (in particular) only ever purported to set out an 'abstract' conception of planning as a rational process of decision-making and action (i.e. a conception of what, hypothetically, it would be for a process of planning to be 'rational'), it is questionable whether it was reasonable to criticise this theory for not advancing any empirical claims. It was a bit like criticising, say, the American John Rawls's famous theory of justice for not being an empirical investigation of the actual administration of justice in the United States. Rawls was not concerned with this, but rather with developing a *philosophical* analysis which would issue in some clarification of our *concept* of justice – some clarification of what it would be for social arrangements to be just. Moreover, such philosophical analysis is far from being 'vacuous' or 'empty'; it is only on the basis of a prior and rigorous analysis of the concept of justice that we can turn to existing empirical reality,[3] and assess whether certain social practices are or are not just. And as with the concept of justice, so too with the concept of rational action outlined by the rational process model of planning.[4]

Nevertheless, it remained true that following a decade of this fundamental conceptual work, by the mid-1970s planners and planning theorists still seemed little further on in developing the kind of empirical understanding of

cities (and the environment generally) which Jacobs, Alexander
rightly called for. It was this gap in planning theory which a later
critics restated, adding to it the need to understand the role a
planning itself within the context of contemporary urban change

100

THE EFFECTS OF BRITISH POST-WAR-PLANNING: 'T CONTAINMENT OF URBAN ENGLAND'

The call by some planning theorists for the empirical investigation of planning and its effects had in some way been anticipated by a major research project undertaken between 1966 and 1971 by a team of researchers led by Peter Hall at the University of Reading. Though there had been studies before to assess the effects of planning, the research undertaken by Hall and his colleagues (1973) and published as *The Containment of Urban England* was by far the most thorough. The team had set itself two tasks. The first was an investigation of urban and metropolitan growth in England from the beginning of the nineteenth century (specifically the 1801 census) to the beginning of the 1970s, concentrating particularly on the interwar years onwards. The second was to focus on the British planning system created after the Second World War and to examine its objectives, operations and effects.

It is impossible to do justice to the wealth of material this contained; it constituted a formidable piece of scholarship and it was widely acclaimed as such – Cullingworth, for example, described it as 'the most significant book to appear on the English planning scene since the Barlow Report'. The second volume of the work is most relevant here, and organised into four main parts: the first examines the concept of planning which underpinned the post-war planning system; the second examines the operation of the system to the end of the 1960s; the third examines the effects of the planning system in practice; and the fourth and final part offers a verdict on the successes and failures of post-war planning. The last two of these parts are summarised here.

In a summary of the findings, Hall (1974) suggested that the post-war planning system had had three main effects down to the 1970s: 'urban containment', 'suburbanisation', and an inflationary effect on land and property prices. The first two were characterised as essentially 'physical', the third as 'economic'. Furthermore, the first effect – urban containment – is arguably the most fundamental in that the two other effects flow from it.

'Urban containment' means simply that the outward spread of urban areas into the surrounding countryside had been 'contained'. As Hall (*ibid*., p. 402) put it: 'the amount of land converted from rural to urban use has been minimised and compacted: urban growth has been *contained*'. This had been achieved by designating green belts around major cities and conurbations; by concentrating new urban development which had to be accommodated beyond green belts in 'substantial pockets' rather than in small-scale or scattered developments, either in coherently planned new towns or in already existing towns and villages; and by accommodating new urban development within cities at high densities, e.g. in high-rise council housing (*ibid*., p. 403).

suburbanisation' – might seem to contradict the first effect of containment, for containing the growth of urban areas seems to imply that suburban sprawl was also contained. However, suburbanisation means something more specific, namely, the increasing separation of homeplaces from workplaces, thus leading to longer journeys to work (i.e. more 'commuting'). This had in part occurred *because* of containment. As new urban development could not be accommodated on the edge of existing cities it had to be located in settlements beyond the green belts. Since most of this new development was for housing, and since, too, many of the residents of this housing continued to work in the major cities, these people obviously faced longer journeys to work. Hence the effect of 'suburbanisation'.

The third main effect – the inflationary effect on land and property prices – was also partly attributed to the policy of urban containment. Restricting the supply of building land when demand was rising, inevitably contributed to a rise in the price of building land and, through this, a rise in property prices. This inflationary effect was fuelled by the planning system itself, for with the nationalisation of development rights developers obviously had to apply to the state (local planning authorities) for permission to develop land, and therefore, as Hall (*ibid.*, p. 403) commented: 'The 1947 system . . . has been directly inflationary by putting a price tag on land zoned for . . . development, and then making it slow or difficult for developers to build on any other land.'

As Hall observed, only the first of these effects was intended by the architects of the planning system. It was one of the prime objectives of post-war planning to prevent urban sprawl but it was not a planning objective to contribute to the lengthening of people's journeys to work; rather, the opposite was intended by making settlements more self-contained. Nor was it the intention to contribute to land and property price inflation. This would inevitably hit hardest the poorer members of society, and the pioneers of town planning wanted to bring about greater equality, with good-quality environments, not exacerbate existing inequalities. And yet, according to Hall and his colleagues, post-war planning had precisely this inegalitarian effect.

As a separate exercise, the authors examined the 'distributive' effects of post-war planning by looking at 'who gained and lost' amongst different social groups.[4] The broad picture which emerged was that, whilst material standards of living had risen absolutely for most people, in terms of *relative* standards (i.e. in terms of material inequalities) the rich had got richer and the poor relatively poorer. Planning had contributed to this effect: 'the effect of planning has been to give more to those that already had most, while taking away from the poor what little they had' (*ibid.*, p. 407).

It was not the intention of those who had designed and administered the post-war planning system in Britain that it should have such inegalitarian effects. As the sober summing up of the distributive effects of post-war planning down to 1970 showed:

> None of this was in the minds of the founding fathers of the planning system. They cared very much for the preservation and the conservation

> of rural England, to be sure. But that was only part of a total package of policies, to be enforced in the interests of all by beneficent central planning. It certainly was not the intention of the founders that people should live cramped lives in homes destined for premature slumdom, far from urban services or jobs; or that city dwellers should live in blank cliffs of flats, far from the ground, without access to play-space for their children. Somewhere along the way, a great ideal was lost, a system distorted and the great mass of people betrayed. (Hall *et al.*, 1973, Vol. 2, p. 433)

But was it true that the planning system was responsible for exacerbating existing inequalities? Notwithstanding the thoroughness of the work undertaken by Hall and his team, this question turned out to be more debatable than the confident conclusion quoted above implied.

EXPLAINING THE EFFECTS OF PLANNING: ALTERNATIVE 'POLITICAL ECONOMY' PERSPECTIVES

Town planning, urban managers and the political economic context

The suggestion that post-war planning practice had contributed to land and property price inflation and, through this, exacerbated social inequalities presupposed that planning was a major cause of these effects. It presupposed that the outcomes could be largely explained in terms of the state's role in planning. Within a few years of the publication of *The Containment* (Hall *et al.*, 1973) this was seriously challenged by sociologists influenced by Marxist social theory.

Up until the mid-1970s, many urban sociologists who had studied post-war planning had tacitly assumed that the planning system was a (if not *the*) key agent in the process of land development, and hence that planning was largely responsible for producing development and the distributive effects which flowed from it. In the language of social theory, a 'managerialist' theory was assumed in which, planners and other local authority officials such as housing managers were viewed as 'managers' of cities and, therefore, as the main agents responsible for urban development and the associated allocation of resources. As Ray Pahl (1975 edn, p. 270) described it: 'The "pure" managerialist model . . . assumes that control of access to local resources and facilities is held by professional officers of the authority concerned.' In suggesting that planning had been a major cause of the pattern of post-war urban development, *The Containment* had, implicitly, assumed something like a managerialist view of planning.

However, in the 1970s some theorists argued that planners and other government officers were in fact much less significant as agents of urban change than the theory of managerialism presumed. This was chiefly because their powers were heavily circumscribed and constrained by more fundamental and enduring socioeconomic forces and 'structures'. If one wanted to understand why certain kinds of urban development were taking place, one would do well to look beyond the statutory planning of any given city or region. What happened in a given locality was often less the result of local political (including local planning) decisions and more the result of 'deeper' economic and social conditions and forces.[5] As Ray Pahl (*ibid.*, pp. 234–5) – a former adherent to managerialism who learnt this lesson – put it:

> the fundamental error of urban sociology was to look to the city for an understanding of the city. Rather the city should be seen as an arena, an understanding of which helps in the understanding of the overall society which creates it. Thus our questions should not be framed in terms of specifically 'urban' problems or 'urban' processes, as if these could be understood separately and independently of the host society.

The message was that the activity and effects of planning should not be interpreted as if planning was an autonomous activity, operating separately from the rest of the society. In explaining planning one had to 'situate' planning activity within its 'political economic' context, because that context significantly shaped and constrained – and therefore did much to explain – the nature and effectiveness of planning activity.

Town planning and the market: Pickvance's analysis

In liberal capitalist societies such as Britain, a central feature of the political economy of land development is obviously the system of private property rights and a 'free' competitive market in land and development.[6] Thus one way of viewing town planning 'within its political economic context' is to assess its effectiveness in shaping urban development in relation to this market system. An example of this analysis was provided by Chris Pickvance in an article published in 1977. Pickvance (1977, p. 70, cited in Paris, 1982) suggests that, to assess the effectiveness of planning, the 'question we need to examine is to what extent the existence of the system of development plans and development control leads to a different allocation of land from a "free market" or "non-planning" situation'. In other words, does planning make any appreciable difference to the pattern of land development (and its effects) than would otherwise arise under free market capitalism?

Pickvance's answer (*ibid.*, p. 69) to this question is unequivocal:

> According to the conventional interpretation of post-war urban development in Britain physical planning is the determining factor and hence physical planners must shoulder the blame for 'failures' such as 'soulless' housing estates, high-rise flats, or the decline of inner-city areas. The aim of this article is to show that the scope of planning powers and the way they have been exercised are quite inconsistent with this interpretation, and that the determining factor in urban development is the operation of market forces subject to very little constraint.

In support of this, Pickvance (*ibid.*, p. 70) draws attention to the fact that, under the development planning system introduced by the Town and Country

Planning Act 1947 (and perpetuated by the 1968 Act, notwithstanding the two-tier system of development plans introduced by that Act), local planning authorities possess only 'negative' powers 'to *refuse* permission permission for development which does not conform to the plan'. Apart from cases where, with the use of compulsory purchase powers, local authorities undertake development themselves – and such public sector development is clearly limited by the availability of public finance – local planning authorities 'have no "positive" powers to ensure that the developments (industrial estates, housing, etc.) set out in a plan will take place' (*ibid.*). Given this, an important consequence follows which, Pickvance (*ibid.*) says, 'is not generally recognised', namely:

> If the planning powers involved in plan preparation and plan implementation (i.e. 'development control') are essentially powers to prevent rather than powers to initiate, then the actual development which does take place depends on the initiators of development or 'developers' . . . and not solely on the preventers of development, the physical planners.

Of course, there is nothing in principle to stop local planning authorities from making a plan with land-use allocations which would not arise under free market conditions. But as Pickvance points out, since their powers of initiating development and therefore of implementing such a plan are limited, such a plan would turn out to be practically useless, for private sector developers will only come forward to develop land if it is in their interest to do so. Anticipating this, local authorities therefore tend to make plans which mirror what is acceptable to developers – that is, plans which set out a pattern of land uses and development which would be likely to arise in any event in a free competitive land market. 'For example, in city centre business and financial districts most planning authorities would not consider any other sort of development besides offices' (*ibid.*). This tendency to make a plan which reflects market conditions is further compounded by the desire of local authorities to maximise their rate income from land development. Thus 'In city centres it is seen as "illogical" to zone land for uses which are not the most profitable and which do not bring in the highest rates income' (*ibid.*). Given all this, land-use planning in Britain has been in practice largely an exercise in 'trend' planning; that is, planning in which 'the development plan merely reflects market trends in the allocation of land' (*ibid.*, p. 71). And thus we arrive at Pickvance's conclusion that planning 'does not lead to a pattern of land uses different from that which would occur in a non-planning situation' (*ibid.*).

To the extent that Pickvance was correct in his conclusion that it is market forces which are primarily responsible for determining the pattern of land development, and thereby the social or distributive effects of development, then the conclusion reached by Hall and his colleagues (1974, p. 407), that 'the effect of planning has been to give more to those who already had most, while taking away from the poor what little they had' was, at the very least, an overexaggeration. For if Pickvance was right, we should conclude that this distributive effect of post-war urban development was more the result of market forces, not planning.

Urbanisation, planning and Marxist political economy

Pickvance's analysis was singled out for special attention because of its clarity and succinctness. In particular, Pickvance vividly brings out how any assessment of the effectiveness of town planning practice cannot be separated from an examination of the political economic context of the market within which planning operates, and this was a crucial theoretical development in the 1970s.

However, Pickvance does tend to speak as if the market and planning are in opposition, as if what town planning would do if only it had the power would be different from the market. There is something in this; planned green belts, for example, run counter to the pattern of urban development that would likely arise if the market were left purely to its own devices. However, in the 1970s another theoretical perspective came to the fore which argued that, very often, the state itself – and town planning as part of the state – worked *with* the given market system rather than as a countervailing force against it. In other words, the state and planning were intimately bound up with and so *part of* the political economic context of liberal capitalism, rather than standing apart from it. Because of this, the state and planning often tended to work in support of capitalism. This way of viewing planning was derived from a Marxist, or historical materialist, view of political economy.

The fundamental premise of historical materialism is that, because human beings must first and foremost supply their material needs (for food, shelter, etc.) in order to exist, the organisation of production (or what Marx called the 'mode of production') is absolutely basic to the organisation of society in general (Marx and Engels, 1846, Part 1, pp. 48–52; Giddens, 1971, Chap. 3). In particular, production to satisfy material needs does not just depend on certain 'powers' or 'forces' of production because production also has to be *organised* in some way (it is this organisation which constitutes the 'mode' of production). Accordingly, in producing things human beings enter into certain 'social relations of production'. Furthermore, the way production is organised and carried out in a society (i.e. the economic foundation of a society) is not separate from the rest of society. Rather, a system of production requires, and therefore 'determines', certain social rules and laws for it to be maintained, which in turn implies a certain system of powers or a political system. More than that, because anyone born and reared in a society will naturally be conditioned into accepting and playing a part in that society, with its particular system of production and laws, etc., it follows that the very way people think – or their consciousness – will be shaped by the kind of social system they are accustomed to. So as Marx himself put it, in the Preface to his *Critique of Political Economy* (1859, Preface):

> In the social production of their life, men enter into definite relations that are indispensable and independent of their will, relations of production which correspond to a definite stage of development of their material productive forces. The sum total of these relations of production constitutes the economic structure of society, the real basis, on which rises a legal and political superstructure, and to which correspond definite forms

of social consciousness. The mode of production of material life conditions the social, political and intellectual life process in general. It is not the consciousness of men that determines their being, but, on the contrary, their social being that determines their consciousness.

In capitalist societies the 'mode of production' is premissed on the private ownership of the means of the production and exchange; this is a fundamental organising or 'structuring' feature of any capitalist society. According to Marxists, from this much else flows, and one of these is that any government which comes to power is not in a position to decide anew what kind of economic system to create. Rather, it inherits an economic system which has developed and been in place for a very long time. So governments and the state generally tend to assume the role of 'managing and maintaining' the economic system – of acting in ways which support and thus strengthen that system. Because in capitalist society the state tends to do things which are supportive of or 'functional' to capitalism, Marxists speak of the state in a capitalist society as being, literally, a 'capitalist' state (see, e.g. Miliband, 1969).

One of the major theoretical questions which arise from this is whether the 'economic base' of society *determines* the 'legal and political superstructure', and hence how far the actions of the state (including town planning as part of the state's actions) are determined by the 'logic' of the capitalist mode of production. There has been intense debate and disagreement between different 'schools' of Marxism on this. 'Structural' Marxists suggest that the state is so bound up with the economic system of capitalism that its actions are almost always governed by, and so explicable in terms of, that system (e.g. Althusser, 1965; Castells, 1977). 'Class theory' Marxists argue that what the state actually does is not strictly determined by that system but is open to 'capture' by the demands of different social classes. Consequently, when the working class mobilises itself as a united class, it can bring pressure on the state to take measures which are in its interests rather than the interests of the capitalist class (see, e.g. Gough, 1979).[7]

What is important to note here is that, in the 1970s, a significant body of urban theorists drew on this general theoretical perspective of historical materialism and related it to the study of urbanisation and town planning (e.g. Harvey, 1973; Cockburn, 1977; Harloe, 1977; Kirk, 1980; Dear and Scott, 1981; Paris, 1982; Rees and Lambert, 1985).[8] In general, these theorists shared the Marxist view of capitalism as an (imperfectly) integrated economic and social *system*, in which the state and planning were part and parcel. As neither the state in general nor town planning as an arm of the state could just step outside that system, there was a tendency for town planning in practice to act in ways which supported the maintenance of capitalism. This was not, as Pickvance implied, that town planning lacked the power to do otherwise (though this was true). More fundamentally, planning was an integral part of that system. As Dear and Scott (1981, p. 4) expressed it: '*neither urbanisation in general, nor urban planning in particular, constitute independent, self-determinate occurrences*. On the contrary, they are *social events*, embedded within society, and deriving their logic and historical meaning from the general pattern of society as a whole.'

Indeed, the suggestion was that a key factor in the creation of statutory town planning was the need, by the state, to find some mechanism to resolve some of the problems (or 'contradictions', as Marxists were fond of saying) of capitalism, such as inefficient patterns of land use, inadequate physical infrastructure, inadequate housing for the working class, and – more generally – the failure of the 'privatised' market to provide 'public' goods such as a pollution-free environment. As the Fainsteins (Fainstein and Fainstein, 1979, pp. 148–9, cited in Paris, 1982) starkly put it:

> Planning is necessary to the ruling class in order to facilitate [capital] accumulation and maintain social control in the face of class conflict. The modes by which urban planners assist accumulation include the development of physical infrastructure, land aggregation and development, containment of negative environmental externalities, and the maintenance of land values . . . Urban planners specialize in managing the contradictions of capitalism manifested in urban form and spatial development.

Here again is an implication that planning (in capitalist societies) is 'determined' by capitalism (it is 'necessary to the ruling class'), that it necessarily acts in ways which are functional to capitalism. This view was certainly expressed by some Marxist urban theorists in the 1970s. For example, Scott and Roweis (1977, p. 1103, emphasis added), after describing the fact that the functioning of market capitalism is inherently prone to generate all kinds of urban problems which undermine the efficient working of capitalism itself, claim that the state is '*compelled* to intervene in various ways' to resolve these problems and, thereby, 'secure the smooth continuation of capitalist society'. However, not all theorists who adopted an historical materialist perspective took this strong 'deterministic' line. Glen McDougall (1979, p. 375) argued that the introduction of statutory planning in Britain at the beginning of the twentieth century (she was writing about the Housing and Town Planning Act 1909) reflected the 'particular configuration of class demands and interests'. This implied that it was not simply determined by the structural requirements of capitalism. Similarly, Backwell and Dickens (1978, Abstract) claimed that, in relation to the planning system established by the Town and Country Planning Act 1947, 'the particular balance of class-forces and relationships between fractions of capital . . . provide a basis for understanding the origins of this system'. They go on to argue that, to some extent, the 1947 system was a response to working-class demands and interests, not just those of capital (much as Gough, 1979, does in relation to the establishment of the welfare state following the Second World War).

Notwithstanding these differences, Marxist urban theorists agreed about one thing in the 1970s: even if the working class had been successful in securing some gains in the capitalist countries of western Europe and North America after the Second World War, the planning systems established in these countries still reflected the power of the capitalist class. *That*, according to

Marxists, was why the state did not introduce forms of town planning which *replaced* the system of private land ownership and property development (the system, that is, under which the means of producing land development were very largely privately owned and controlled), but instead put in place planning systems which merely *regulated* capitalist land development (see McKay and Cox, 1979, Chap. 2). *That*, too, was why, in practice, the planning system had not generally produced outcomes which were significantly different from what the capitalist land market would probably have produced. The 'effects' Hall and his colleagues (1973) attributed to planning were thus largely attributed by the Marxist theorists of the 1970s to the prevailing socioeconomic system of capitalism.

CONCLUSION: TOWN PLANNING AT THE MACRO AND MICRO SCALES

What were – and for that matter (for the question is still pertinent today) what *are* – the effects of town and country planning? This is the main question which preoccupied planning theorists in the late 1970s and early 1980s. What began as an apparently straightforward story of cause and effect – with the planning system being seen as a prime cause of the pattern of post-war development – came to be seen, by the late 1970s, as much more complex, and hence a matter of theoretical controversy. In this chapter I have tried to set down the main outlines of that theoretical debate, describing the different theoretical positions which were held.

One of the theoretical perspectives – the 'political economy' perspective on planning – has had a lasting influence on the way planning theorists now think about planning (see Chapter 7). It is now generally accepted that one cannot investigate the effects of the planning system independent of its political economic context, and that the market system of land development in particular plays a crucial role in determining the outcomes of planning practice. However, the theoretical disputes of the 1970s and 1980s did not settle once and for all the problems of understanding the role and effectiveness of planning. This debate over the nature and determinants of contemporary urbanisation, and the role of the planning system, has continued.

In retrospect, just as the theoretical position of urban managerialism had exaggerated the power of the planning system, so some Marxist writing in the 1970s tended to exaggerate the weakness and thus ineffectiveness of the planning system in comparison with the capitalist market system. A more considered reading of the political economy of planning suggests that the picture may be more complex than either side of that theoretical divide acknowledged in the 1970s. In particular, it may be more fruitful to view the process of urban development as being shaped by a variety of factors and agencies (of which statutory planning is one) whose respective 'weights' may vary in different circumstances (cf. Ball, 1983). What Pickvance calls 'market forces' may generally be the most 'weighty' of these forces, but this needs further comment.

In different economic circumstances the role planning can play in shaping urban development can be less, or more, strong. Thus when a local economy is buoyant and there is strong pressure from the private sector for development, local planning authorities can exert greater pressure on developers to conform to planning ideals which the market would otherwise not realise, e.g. in terms of the location and density of development; the design of buildings, spaces, and landscape; the provision of public spaces; the provision of needed infrastructure. By contrast, in areas of economic stagnation and decline, any would-be developer is obviously in a stronger position to dictate the nature of the development although even here planning authorities may have some room for manouvre.

One thing characterised the political-economic theory which emerged in the 1970s – it concentrated on the role and effectiveness of town planning at the 'macro' scale – the scale of the operation of the (British) planning system *in general* or *as a whole*. However, this is not the only possible way of studying the effects of town planning actions. It is possible, for example, to examine the effects of planning in a much more focused or particular way at the 'micro' scale, by examining the effects of particular plans or policies in particular settings. As Eric Reade (1987, p. 70) has pointed out, examining the effects of particular policies is equivalent to the *monitoring* of policies with a view, if necessary, to reviewing or revising them if they are not having the desired effects. As Reade also observed, when he was writing in the mid-1980s, up to that date there had been remarkably little systematic empirical research into the effects of specific planning policies. Reade offered an explanation for this dearth of focused research, and set down some methodological criteria which should guide such research when it is undertaken.

As Reade points out, in many areas of public policy there is a tradition of carefully researching the effects of specific policies in order to test their effectiveness. Thus in the field of government economic policy-making, specific policies such as alterations in interest rates or the introduction of a new tax will be carefully scrutinised and monitored and then reviewed in the light of what seem to be their effects. Isolating the effects of a policy in this way is extremely difficult and is always open to uncertainty. Reade (*ibid*., p. 69) acknowledges this:

> To discover these effects . . . is far from easy; causation is not so easily established. The web of governmental activity is so wide and so dense, and so intricately enmeshed with the economy and with social institutions more generally, that it must often seem difficult to believe that any outcome can be explained, other than in terms of virtually everything else.

However, if literally no light can be shed on the likely effects of a specific policy this obviously cuts away the justification of that policy, however well meaning it may be. For if we have no idea of the effects of a policy, it may have effects contrary to the ones desired. It is therefore a precondition of any rational appraisal of public policy that there be an *attempt* to identify the likely effects of policies, however difficult this may be.

In spite of this, Reade argues, there has emerged in Britain a tradition of town planning which has compounded the difficulties of empirically examining the effects of specific policies. This is the tradition of 'Utopian comprehensiveness' which has sought to conceive of the ideal city or settlement *as a whole*. From this tradition developed the practice of planning settlements 'holistically' (or 'comprehensively') so that 'comprehensive' development plans were produced for towns in which whole clusters of policies were welded together as a 'package'. In Popper's terms (1957, Chap. 3), town planning thought and practice has been characterised by 'holistic' or 'Utopian' social engineering. And as Reade (1987, p. 84) points out, the problem with this is that: 'If too many policy changes are attempted simultaneously . . . [it is] impossible to know which of the observed effects to ascribe to which of the policy changes'.

In contrast to holistic or Utopian social engineering, Popper proposed that public policy should proceed by what he termed 'piecemeal' social engineering, in which policy changes are introduced in a cautious, piecemeal fashion so that the effects of individual policies can be more readily monitored. As noted above, even the task of understanding the effects of individual policies is complex enough, but if policy-making and the examination of policy effects are approached in this Popperian piecemeal way, there is at least a greater chance of identifying the effects of public policies. Or so at least Popper and Reade argue.

Reade's explanation of the dearth of research into the effects of planning policies at the micro level leads him to specify what in his view needs to be done to rectify this deficiency. He suggests three 'methodological criteria' for researching the effects of planning policies. First, it is vital to distinguish one policy from another and, relatedly, to be clear what a given policy is designed to achieve (i.e. what its intended effects are supposed to be). Secondly, after a planning policy has been in place for some time an attempt can be made to identify what *physical* effects it has had on urban development to see whether these are the effects it was designed to realise. For example, we could examine whether a policy has helped to cause (or prevent or restrict) new development in certain locations, and if so, what kind of development and how much, at what density, with what design etc. Thirdly, one should then try to identify what '*social*' effects these policies have had or are having by examining how different social groups have been affected.

As Reade acknowledges, there is nothing very startling about these guidelines. Nevertheless, the capacity to adopt these methodological principles does depend on town planning being approached in a rather different way from how it has been practised hitherto. In particular, it would require an approach to planning which seeks to devise particular policies to address particular problems rather than one which generates a whole set of policies in a single comprehensive plan. In Popper's terms, it would require a 'piecemeal' rather than a 'Utopian' approach to town planning.

NOTES

1. There was one interesting difference between the criticisms of the earlier generation of critics, such as Jacobs and Alexander, and those of the later generation, such as Scott and Roweis, Camhis and Thomas. Jacobs and Alexander chided planners for their lack of understanding of *cities* (i.e. the objects they were hoping to plan), whereas Scott and Roweis, and Camhis and Thomas criticised (procedural) planning theory for its lack of understanding of (actually existing) *planning*. Implicit in each was therefore the call for two different kinds of theory relevant to planning, one about cities and the other about planning itself. However, both generations were united in calling for planning to be informed by much greater understanding of empirical reality – and thus for planning theory to be grounded in the empirical investigation of the real world.
2. Though for one fruitful investigation, in this case of the functioning of retail and market centres, see Berry (1967).
3. For a defence of the rational process theory of planning in these terms, see Taylor (1984; 1985).
4. The conceptual basis of the authors' categorisation of social groups for this exercise is unclear. Thus some groups seem to be identified in terms of their *geographical location* (e.g. 'rural dwellers', 'new suburbanites'); others in terms of their *socioeconomic position* (e.g. the 'more affluent', the 'less affluent home owner', the 'tenant in public housing'). See Hall *et al.* (1973, Vol. 2, pp. 405–8; Hall (1974, p. 406).
5. One general account of this theoretical position was succinctly stated by Steven Lukes (1974), in terms of a theory of power.
6. 'Free' is in quotation marks because, in western liberal societies, no market operates completely free of state regulation. Thus all market transactions are underwritten by laws concerning ownership rights, fair trading, contracts, fraud and so on. In this respect it has always been false to say that liberalism supports *laissez-faire* or free markets in *that* (extreme) sense.
7. Marxists of this persuasion are fond of citing Karl Marx's famous statement that: 'Men make their own history, but they do not make it just as they please; they do not make it under circumstances chosen by themselves, but under circumstances directly encountered, given, and transmitted from the past' (Marx, 1869, p. 360).
8. Of these the key path-breaking text in the English-speaking world was David Harvey's *Social Justice and the City* (published, interestingly, in the same year as *The Containment of Urban England* – 1973). The reason I do not deal with Harvey's book at greater length is that, for all its virtues, it does not provide a specific analysis of the role and effectiveness of *town planning* from a Marxist perspective.

7

Rational planning and implementation

INTRODUCTION

In Chapter 6 we saw how, in the 1970s, a 'second wave' of criticisms of post-war planning theory emerged, directed this time mainly against the rational process view of planning, or 'procedural planning theory'. There were two main criticisms in this second wave, and one of these was covered in Chapter 6, namely, the criticism that procedural planning theory was 'contentless' or 'empty' because it was not based on the empirical study of actual planning practice and therefore said nothing about it.

The second criticism of post-war planning theory introduced briefly in Chapter 6 was that, in some way, the rational planning model distracted attention away from the crucial question of how plans and policies were implemented, if they were implemented at all.[1] This problem in fact predated the emergence of the rational planning model of the 1960s. Thus in the early post-war era, the activity of town planning in the forward-planning sections of local planning authorities was considered to be a creative exercise of designing and *making plans* ('master' plans), and little thought was given to how these plans might be implemented. Most town planners received no training in, and therefore lacked any practical skills for, implementation. Even the fact that planning involved a political process in which well intentioned plans might be thwarted by disagreement was not something many planners took much notice of until the 1960s. Consequently, the plan chests of many local planning authorities were stuffed with 'bottom-drawer' plans that had not been through the practical, messier business of implementation.

Protagonists of the rational process model also gave little attention to how plans and policies were, or might be, implemented. However, the rational process model did not necessarily imply that implementation was glossed over or ignored. Implementation was explicitly identified as part of any rational planning process (see Chapter 4, Figure 4.2), and so the rational process model should always have been seen as a model of rational *action*.

However, when the rational model came to prominence, most attention was paid to the 'early' policy-formation stages of the process and little to the 'later' stages of implementation and monitoring. There was thus considerable discussion of planning goals and objectives, including such empty debates as whether planning was best conceived as an exercise in solving problems or achieving goals (see, e.g. Faludi, 1971; Needham, 1971; Gutch, 1972). Likewise, much attention was devoted to plan and policy-*making* (how best to generate and specify alternatives), and to policy and plan *evaluation* (there was a debate about Morris Hill's goals achievement method of evaluation as compared with cost-benefit analysis or Lichfield's planning balance sheet[2]). It is therefore significant that, in McLoughlin's seminal textbook on the systems approach to planning (1969), in which the chapters are ordered in terms of the various stages of the rational planning process, only 17 of the relevant 195 pages are devoted to plan implementation. It is therefore not surprising that the rational process model was generally described as a model of rational *decision-making*, rather than a model of rational *action*.

This criticism was not confined to town planning; it was a weakness of public policy generally. The path-breaking text was Pressman and Wildavsky's *Implementation* (1973), which was an examination of policy-making in the area of economic development by the US government. Once again we see how an issue of concern to policy analysts and theorists generally came to be adopted in the specific field of public policy concerned with town planning.

Although this concern over implementation was prompted by failures in planning practice, and was therefore an intensely practical concern, considered reflection on the problems of implementation naturally spawned a literature in its own right and so became a matter of theoretical attention. Early work on implementation threw up two issues. First, some of the early writings suggested that a proper engagement with implementation required attending to action more than plan or policy-making. This naturally provoked some debate about the relationship between policy and action. Secondly, early theorists soon recognised that effective implementation required the interpersonal skills of communication and negotiation. In the 1980s and 1990s therefore a new planning theory came to prominence which viewed planning as an exercise in 'communicative action'.

In the town planning systems of all western liberal democracies since 1945, public sector town planning has coexisted with a capitalist land market. Implementing public policies and plans therefore requires planners who understand and are prepared to work with the initiators of development, including, obviously, private sector developers. The kind of theory concerned with understanding the political economic context of town planning (described in Chapter 6) in which the market's strong role was emphasised, ties in with the issue of implementation. In the concluding section I discuss the connection between political economic theory of planning and theory about planning implementation.

THE EMERGENCE OF IMPLEMENTATION THEORY

The origins of the critique of the rational process theory of planning go back to Lindblom's criticism of 'rational comprehensive' planning (Lindblom, 1959; see also Chapter 4). Central to Lindblom's questioning was a critique of the ideal of comprehensiveness which, Lindblom contended, could never be achieved in practice (*ibid.*, p. 160, cited in in Faludi, 1973a). Lindblom's alternative model of 'disjointed incremental' planning was thus advanced as a more realistic account of what the process of planning was like, and could only be like, in practice. Although Lindblom did not address the issue of implementation specifically, implicit in his more pragmatic model of planning was an appreciation of the practical difficulties of implementing policies in the real world.

Perhaps the first critique of the rational model of planning that was linked explicitly to a concern with implementation was that advanced by the American John Friedmann (1969). The opening sentence of Friedmann's article suggested that there had been a tendency to separate the activity of making plans from the business of implementing them: 'The idea that planning and implementation are two distinct and separable activities dies hard' (*ibid.*, p. 311). Friedmann goes on to quote from the report of a conference of urban planners in 1967 in which it had been said that 'Society has been planning too much and experiencing far too little application of plans – there is yet to be created a climate for the acceptance of plans. The problem of implementation is a crucial one' (*ibid.*).

Friedmann's critique of planning theory for ignoring the problem of implementation – or, as he also put it, the problem of 'action' – was directed particularly at the rational model of planning. The problem as he saw it was that it was advanced as essentially a theory of how best to make *decisions*, and this led planners to ignore *action*.[3] As he put it, somewhat tendentiously: 'The problem is no longer to make decisions "more rational", but how to improve the *quality of the action*' (*ibid.*). This is tendentious because it implies a dichotomy between rationality and action where in fact the challenge is not to engage in action (or implementation) which is non-rational – not, that is, to focus on action *at the expense* of reason – but rather to secure effective, but still rational, action. The problem with the rational process model was not therefore the model's concern with rationality as such but with the particular way in which the process of rational planning was conceived and represented.[4]

Friedmann focused on the fact that the rational process model of planning is typically presented (as in Figure 4.2) as a series of linear steps or stages which are to be approached one after the other. If a rational approach to planning is conceived in this way, then – Friedmann claimed – there was a tendency to separate the task of plan-making from that of implementation. This was because the task of plan-making was shown as a separate stage of the process, and one which came *before* that of implemention. Planners would therefore attend first to the task of making plans and only later and separately to the problem of how to put those plans into effect. Friedmann (*ibid.*, p. 312) describes the dangers of this:

> According to the classical *decision model* of the planning process, four distinct steps are involved: (1) preparation of alternative plans by planners; (2) adoption of one of these plans by deciders; (3) implementation of the chosen plan by administrators; and (4) recycling information concerning the results of implementation to planners who use this information to

> revise the current plan. It is this model which has caused the difficulties alluded to earlier. Plans are made, but the deciders proceed to actions which are not in accord with any the plan proposed.

But it is worth asking what, exactly, is wrong with this linear way of conceiving rational planning. After all, isn't the problem of implementation precisely one of how to put into effect previously decided plans and policies? Logically, the problem of implementation only arises when you first have something to implement. And so it would seem that one has no choice but to *first* make policy and then implement it.

On page 119 I discuss the work of some theorists who have questioned this logic – who question the idea that implementation necessarily involves putting some policy into effect, and so dispute the above statement that there has to be some prior policy one wishes to enact for the task of implementation to arise. But even if we accept that the making of a policy or plan must precede implementation, there still remains a problem with conceiving the process of rational planning in a linear, step-by-step fashion. If one first makes plans *without at the same time considering the problem or 'stage' of implementation*, one is liable to make plans which cannot be implemented. For example, insufficient attention is given to whether the desired plans are feasible or to whether there are resources available to enact them and so on. Or one is likely to give insufficient attention to the question of whether others (e.g. land developers) are willing to implement one's plans, and implementation depends on the willing co-operation of at least some others. This is certainly the case in the planning systems of liberal capitalist societies, such as Britain and the USA, where the implementation of development proposals generally depends crucially on their acceptability to private sector developers. In these respects, therefore, implementation is not something which can sensibly be left 'until later' or to a 'later stage' of the planning process. Rather, the task of implementing plans needs to be considered *at the same time* as plans are made.

As with plans and policies, so likewise with the aims and objectives of a plan which, in standard accounts of the rational planning process, come 'first' in the process. There is no point in enunciating fine-sounding aims and objectives which have no chance of being realised – which are not 'realistic'. So here, too, in defining goals and objectives it is important to consider *simultaneously* questions of implementation.

It is in the light of these considerations that the standard diagrams of the rational planning process which separate out the different stages of the process and order them in a linear step-by-step fashion, are open to the criticism that they conveyed a misleading picture of rational planning and action. While one can conceptually distinguish between different intellectual tasks involved in making rational decisions and engaging in rational action – identifying aims, formulating alternatives, evaluating alternatives – it is dangerous to view and undertake these different tasks as separate 'stages' in a linear process. As Friedmann (*ibid.*) put it, in advocating an alternative 'action-centred' model of rational planning:

> By contrast, the new approach which may be called the *action planning model* fused action and planning into a single operation so that the conceptual distinctions of planning–decision–implementation–recycling are washed out. This model is close to observable reality. In most situations, it is extremely difficult empirically to isolate the four phases of the classical model, particularly the critical step of decision-making. An action will include deliberation and choice as pervasive, but these are not to be identified as distinctive phases *prior* to action; they are – inseparably – a part of it.

Again we must be careful, for Friedmann once more overstates his point. It is not that the conceptual distinctions between different components of rational action are, or ought to be, 'washed out'. For the making of such distinctions is vital if one is to be clear about the different aspects of engaging in rational action. Thus it is vital to be aware of the need to think about and specify what one is aiming to achieve, to consider alternatives and carefully evaluate them etc. The point is that, when dealing with any one of these 'stages' there are serious dangers if we forget about the other 'stages'. In particular, if we engage in plan-making 'on its own' there is the danger we shall ignore necessary questions about implementation. This is the main point Friedmann was driving at, and it represented a serious criticism of the rational process view of planning as it was conceived and put forward in the 1960s and 1970s.

The suggestion that there might be a tendency to make plans and decisions without simultaneously considering the problems of implementation was not confined to the specific field of town planning. These criticisms were made of public policy-making generally, as was made clear in Pressman and Wildavsky's seminal book *Implementation*, published in 1973. Subtitled 'How great expectations in Washington are dashed in Oakland', the book described how a new agency established by the US Congress in 1965, the Economic Development Agency (EDA), devised a programme to provide stable employment to disadvantaged minorities living in US cities, and chose Oakland as an experiment to spearhead the programme.

One of the features of the programme was that it was not initially hampered by political or economic difficulties. As the authors say: 'Some programs are aborted because political agreement cannot be obtained. Others languish because funds cannot be secured. Still others die because the initial agreement of local officials or private concerns is not forthcoming' (*ibid*., p. xx). But this was not the case with the EDA's Oakland initiative. Here the programme was agreed by the Federal government and the necessary funds were made available, both the city officials of Oakland and local employers approved of the plan, and the programme was launched with widespread publicity. Yet in spite of all this the programme foundered and, whilst some gains were achieved, they were slight compared with the high hopes which accompanied its inception. As Pressman and Wildavsky wrote (*ibid*.): 'construction has only been partially completed, business loans have died entirely, and the results in terms of minority employment are meager and disappointing.' So what went wrong?

Pressman and Wildavsky (*ibid.*) show that the Oakland Project foundered largely as a result of obstacles of 'a prosaic and everyday character'. For example, two major publicly funded infrastructure projects, for a new marine terminal and an aircraft maintenance hangar, were both beset by numerous unforeseen difficulties and delays: the need to secure agreements and clearances from various agents whose co-operation was essential to the realisation of the projects; the need to obtain interim funding to initiate work before Federal funds came on stream; unanticipated design difficulties associated with the projects, such as the quality of the filling material for the marine terminal and adequate fire protection for the aircraft hangar; the escalation of costs owing to these delays and difficulties which required the securing of further agreements for additional public funds before the projects could proceed; and so on.

The moral was clear. For a plan or project to be successfully realised it is necessary to anticipate and resolve the problems which might frustrate its implementation. Given that the actual success of a plan depends on whether or not it is successfully implemented, it follows that just as much attention needs to be given to the task of implementation as is given to the formulation of a plan. Or, returning to the discussion of Friedmann, implementation needs to be considered at the same time as and not after the stage at which plans and policies are formulated. Otherwise 'great expectations' are likely to be dashed.

Pressman and Wildavsky (*ibid.*, Chaps. 5 and 6) drew out some of the lessons to be learnt from their research: thinking through the complexities of implementing what may at first appear to be quite simple actions; recognising that the realisation of most projects involves the participation of a number of actors, from which it follows that contact must be made with these to secure their agreement in the first place; ensuring that an effective team is put together to work on a plan or project, and that key responsibilities for delivering aspects of the programme are made clear; anticipating and preparing for changes in key personnel during the lifetime of the project (one factor affecting the difficulties of the Oakland project were changes in key personnel at a crucial stage in the life of the programme); costing proposals, especially where they depend on public funding or the agreement of funding agencies; and so on.

Most of these are fairly obvious, when one reflects on what is necessary to carry through a planned action to successful completion, and on what can impede or thwart the realisation of a plan or policy. Yet things are 'obvious' only when one pays attention to the practical business of implementation. And because so few planning theorists had attended to implementation, the 'obviousness' of these considerations had hitherto eluded most planning theorists.

One of the most important things Pressman and Wildavsky's work shows is that implementation of public policy rarely depends on the actions of the relevant government department or planning authority alone. Social action rarely depends solely on a single actor but usually involves and requires the co-operation of different actors (recognising this, Pressman and Wildavsky devoted a whole chapter of their book to the problems of 'joint action'). Moreover, these other actors have 'private' goals of their own, some of which do not coincide with those of public policy-making authorities. This is the case with town planning in liberal capitalist societies, where the realisation of development proposals in public plans depends on private developers to carry out the development, and private developers will typically only do this when the development will profit them.

To become effective implementers, public authority planners and policy-makers need to become skilled in three tasks. First, they must be able to identify the other actors necessary for or relevant to the implementation of a plan or policy. Secondly, they must establish contacts with these actors. Thirdly, because these other agents have their own objectives which do not always coincide with those of the public authority, planners and policy-makers must acquire the skill of negotiating. Taken together, all these tasks require interpersonal skills; in short, effective implementation requires planners who are skilled at *contacting*, *communicating* and *negotiating* with others.

If town planning is to 'have some effect' on the world, then it has to be viewed as centrally about *action*. Viewed in this way, in addition to formulating or making elaborate plans or policies, the 'action-oriented' planner needs to develop interpersonal skills which are necessary for effective action. On this, it interesting to return to Friedmann (1969, pp. 316–17), for he was remarkably prescient in seeing, and advocating, the link between this kind of 'action-oriented' planning and the interpersonal skills required for it:

> Where action and planning are fused, the role of the planner changes fundamentally. The planner formed in the image of the classical model was primarily a technician, an analyst and a model-maker. Relatively isolated from the vital forces of change in society, he saw the world in symbolic abstractions such as figures, graphs, charts, and maps . . . But a new breed of action planners, oriented to a different professional image, is moving forward . . . To be involved in action is to interact with others who contribute skills and knowledge that are different from those of planners – such as politicians, administrators, influential persons, 'gatekeepers', representatives of interest groups, technical staffs from competitive institutions, and many more . . . In action-planning, then, the planner moves to the foreground as a person and autonomous agent. His success will in large measure depend on his skill in managing interpersonal relations . . . the planner has to learn to live with conflict . . . and to exploit conflicting forces for constructive action. Only rarely will the planner have his way; he will have to bargain, compromise, and learn to accept defeat without being crushed by it.

This statement raises two important issues about the nature of implementation. First, Friedmann's suggestion that a new concept of planning – 'action-planning' – should supersede the 'classical' model of planning as policy and plan-making. However, Friedmann's suggestion naturally raised the question of what should be the relationship between policy or plan-making on the one hand, and action or implementation on the other. To ask this question is simultaneously to ask how the activity of implementation itself should be conceived. Consider two alternative ways of viewing the relationship between policy and action. One is that planning necessarily involves *first* making plans or policies (albeit feasible and therefore implementable ones) and *then* enacting or implementing them. If one takes this view one is simultaneously adopting a concept of implementation which sees it as 'putting policy into effect'; policy comes first and implementation follows, seeking to put a given policy into effect. A second alternative view – which seems to be the one Friedmann is advancing when he talks of 'fusing' action and planning – would not separate planning (or policy-making) and action in a sequential fashion but rather would see planning as *being* a form of action and hence 'implementation' through and through. The 'action perspective' implies that such goals or policies need to be adapted to the circumstances planners find themselves in, so that they cannot be disassociated from the realm of action. This perspective allows for plans and policies to be sometimes created 'on the spot', because changes in the real world may have taken place to render policies redundant or inappropriate, or because there are no existing policies in place to deal with the circumstances planners find they are facing, etc.[5] Therefore, far from policy preceding action, policy often *follows* or is a *response* to action. It may therefore be more appropriate to conceive of implementation, not as 'putting policy into effect', but rather as about action 'to begin with'. Under this conception of implementation, policy making would be seen less as something which precedes action, and more as *part of action* (or part of implementation). This issue of the relationship between policy and action, and with it the question of how we should best conceive implementation, was debated by some implementation theorists in the 1980s, and in the next section I shall discuss this debate.

The second issue is that Friedmann's view, that the 'action-oriented' planner would be one who is skilled at 'managing interpersonal relations', echoes the conclusion reached by Pressman and Wildavsky that implementation requires interpersonal skills of 'networking', communication and negotiation. This became an important area of concern for some planning theorists through the 1980s – so much so that, by the 1990s, one version of communication theory has come to be seen as the now dominant 'paradigm' of planning theory (Innes, 1995) (see p. 122ff).

IMPLEMENTATION THEORY AND THE RELATIONSHIP BETWEEN POLICY AND ACTION

In the previous section I argued that the kernel of Friedmann's criticism of traditional approaches to planning and policy-making was the separation of policy and plan making – or indeed of any 'stage' of the rational planning process – from considerations of implementation. Friedmann's point was that successful planning required considering the likely problems of implementation *at the same time* as plans and policies are being prepared. 'Effective plan implementation begins at the early stages of plan preparation . . . Plan formulation is not an autonomous phase in planning' (Friedmann, 1969, p. 312).

Even if considerations of implementation were to be treated concurrently with policy and plan preparation (or goal formulation, plan evaluation, etc.), this way of dealing with implementation still failed to satisfy some implementation theorists. It still remained embedded in a model which viewed the planning process as a linear series of discrete stages in which plans or policies are first formulated and *then* put into effect. Thus Barrett and Fudge (1981, p. 10, emphasis added), in reviewing the literature on implementation, criticise Pressman and Wildavsky because 'they assume a series of logical steps – a progression from intention through decision to action – and clearly see implementation starting where policy stops . . . they see implementation as a process of *putting policy* (or in their case, programmes) *into effect*'.

Barrett and Fudge characterise this way of viewing implementation as the 'policy-centred approach', in which 'policy is the starting point, the trigger for action, and implementation a logical step-by-step progression from policy intention to action' (*ibid.*, p. 12). They also add that: 'This approach might be defined as "the policy-makers' perspective", since it represents what policy-makers are trying to do to put policy into effect' (*ibid.*).

According to Barrett and Fudge (*ibid.*), what is wrong with this way of conceiving implementation is that 'it assumes that policy comes from the top and is the starting point for implementation and action'. However,

> This . . . is not necessarily the case: policy may be a response to pressures and problems experienced on the ground. Equally, policy may be developed from specific innovations, that is, action precedes policy. Not all action relates to a specific or explicit policy. The hierarchical view of implementation also implies that implementers are agents for policy-makers and are therefore in a *compliant* relationship to policy-makers. But in many instances – especially in the public policy field – those upon whom action depends are *not* in any hierarchical association with those making policy. By definition, public policy is often aimed at directing or intervening in the activities of private interests and agencies. Implementation agencies will thus, in many instances, be autonomous or semi-autonomous, with their own interests and priorities to pursue and their own policy-making role.

Given this, they counterpose what they call an 'action perspective' against the aforementioned 'policy-centred' or 'policy-makers' perspective'. This they explain (*ibid.*, pp. 12–13) as follows:

> it is essential to look at implementation not solely in terms of putting policy into effect, but also in terms of observing what actually happens or gets done and seeking to understand how and why. This kind of action perspective takes 'what is done' as central, focuses attention on the behaviour or actions of groups and individuals and the determinants of that behaviour, and seeks to examine the degree to which action relates to policy, rather than assuming it to follow from policy. From this perspective, implementation (or action) may be regarded as a series of *responses*:

> to ideological commitment, to environmental pressures, or to pressures from other agencies (groups) seeking to influence or control action.

It should be clear from this that Barrett and Fudge's 'action perspective' has much in common with Friedmann's 'action-planning', in which (according to Friedmann, 1969, p. 316) 'action and planning are fused'. However, more explicitly than Friedmann, Barrett and Fudge dispute the assumption that policy necessarily precedes action, and hence the idea that implementation has to be conceived as 'putting policy into effect'. What should we make of this more radical development of Friedmann's original notion of action planning?

Barrett and Fudge's suggestion can be regarded as a distinctive and significant advance in thinking about implementation. Under the 'policy-centred' view they criticise (i.e. the view which sees implementation as 'putting policy into effect'), even when policy is framed with considerations of implementation in mind our attention may still be primarily focused on the business of *formulating* plans and policies, and hence on *statements, documents, plans,* – the 'paperwork' of policy-making. In this way, we may fail to attend adequately to the sphere of practical action. In preparing plans or policies, we can simultaneously anticipate and address questions of implementation, such as the feasibility of a policy, the resources needed to realise it, its likely attractiveness to other actors who will have to be involved in its enactment. Certainly this is an advance on the 'master planning' approach to plan or policy making. Yet the 'policy-centred' view remains a conception of implementation which is tied to the making of policy statements, the production of policy documents and so on. In attending to these things one is not attending to the world of action which must be confronted if one is to succeed in changing reality; after all, policy documents and statements are not *themselves* reality-changing actions. To concentrate on policy (and then on putting policy into effect) is not to concentrate on action, and implementation must be about action. To address the real business of implementation, one has to develop an understanding of, and engage with, the world of action proper rather than the world of policy statements and documents. This was the force of Barrett and Fudge's action-perspective.

But it is worth noting two points that could be raised in criticism of Barrett and Fudge's account. Even if we allow that policies may sometimes have to be made up in the course of action as 'responses' to action, or that given policies may have to be flexibly adapted to unforseen 'action' situations, it can still be argued that implementation involves, unavoidably, 'putting policy (of *some* kind) into effect'. In this regard, we should note that Barrett and Fudge's argument against this idea trades on a couple of questionable conceptual distinctions. The first concerns their observation that, in reality policy is sometimes a response to action rather than the thing which initiates action by being 'put into effect'. However, this sets up a false dichotomy. For it may be true that a given policy P is created in response to action, but this remains compatible with the idea that the *implementation* of P still involves trying, in some way, to put P *into effect*. The suggestion that policy is sometimes a response to action is not an *alternative* to the suggestion that implementation involves putting policy into effect. The former view is about how policy *originates*, and thus it is arguably not about implementation at all; the latter view is clearly about implementation – *however* policy originates, implementation is about its *enactment*. The observation that policy is sometimes a response to action is something of a truism for plans and policies are *usually* a response to some actions taken by others.[6] Policy is never devised in a complete vacuum or 'just for the sake of it'; rather, policy is usually devised in response to some problem, and – with the exception of natural disasters – the problems we devise policies to solve are the results of prior human actions.

The second questionable point Barrett and Fudge make is when they speak of 'implementation (or action)' as if these two things were one and the same thing. But they are not necessarily the same thing. 'Implementation' is only a *specific kind of action* distinct from all the other kinds of actions we can engage in. Barrett and Fudge fail to distinguish between action in general (including actions within the policy-maker's field of action) and the action of implementation in particular. Seeking to understand 'action' in the general sense recommended by Barrett and Fudge is not necessarily the same as addressing the problem of implementation. 'Implementation' necessarily involves putting some kind of policy into effect. To engage in implementation entails implementing *something*, and that 'something' has to be a policy (or plan, goal, etc.) of some kind. *Ergo*: implementation = 'putting policy into effect'.

If these two counterarguments hold, then the proposed distinction between 'policy-centred' and 'action-centred' views of implementation was not as clear as Barrett and Fudge implied. This is not to deny Barrett and Fudge's (and before them, Friedmann's) central thesis that effective implementation requires an understanding of the field of action. Nor is it to deny that being an effective implementer requires action-oriented skills. It remains, however, that implementation is about implementing something, and that something is some kind of policy or goal. If this were not true then, in engaging in action, planners and policy-makers would have no notion of what they are engaging in action *for*, or what they are *planning* for.

These conclusions mirror those of Eugene Bardach (1977). In Bardach's view, successful implementation requires being effective at both policy-making and action so that, for him, there is no dichotomy between 'policy and action'. Bardach asserts (*ibid*., p. 5) that the 'most important approach to solving, or at least ameliorating' the implementation problem 'is to design policies and programs that in their basic conception are able to withstand buffeting by a constantly shifting set of political and social pressures during the implementation phase'. Being able to do this, however, requires understanding and, where possible, working with other agents including some who may be antipathetic to the objectives of a public authority. Moreover, Bardach emphasises that even the most robust policy which is 'well designed to survive the implementation process' may go awry because 'many implementation problems are inherently unpredictable' (*ibid*.). Successful implementation of public policy requires individuals who are effective at 'fixing' things if policies do go awry (*ibid*., Chap. 11). Like Barrett and Fudge, therefore, Bardach also draws attention to the need for effective implementers to be effective in the world of action. Yet in doing so he does not lose sight of the fact that the problem of implementation remains, ultimately, the problem of implementing some policy or plan.

PLANNING AS 'COMMUNICATIVE ACTION'

The early work of Pressman and Wildavsky on implementation, and of John Friedmann on 'action-oriented' planning, drew attention to the importance of interpersonal skills of communication and negotiation for the effective implementation of policies and plans. These interpersonal skills continued to be regarded as central to those planning theorists who, through the 1980s, continued to develop an 'action-centred' view of planning.

So central were the interpersonal skills planners should develop that, by the early 1990s, a whole new theory of planning came to be articulated around the idea of planning as a process of communication and negotiation. Thus Sager (1994) spoke of a new 'communicative planning theory' whilst Innes (1995) pointed to an emerging 'paradigm' in planning theory concerned with 'communicative action and interactive practice'.

The concern with 'communication' was not entirely new in town planning. In the days of master planning, great attention was given to the presentation of plans on the grounds that planners should communicate their proposals clearly and attractively. However, the emphasis was very much on the presentation of plans, and in this was an assumption that communication was primarily a oneway process *from* the planner *to* politicians and the public. Little attention was paid to communication as an *interpersonal* activity involving dialogue, debate and negotiation. It was the interpersonal nature of communication that was emphasised by the communicative planning theory which came to prominence in the 1990s.

It is important to note that (rather as we saw with 'implementation' and 'action' earlier) 'communication' and 'negotiation' are not necessarily the same thing. *Communication* refers to the business of communicating *in general*, whereas *negotiation* is a *specific kind of interpersonal communication*. Thus, someone could be generally an effective communicator but not an effective negotiator. The converse, however, does not apply: it is difficult to imagine someone being an effective negotiator who is not also an effective communicator. In other words, effective communication would seem to be a necessary but not a sufficient condition of effective negotiation. If the two activities are distinct, this implies there are distinct skills associated with each. So given that effective implementation requires skills of communication *and* negotiation, one would expect planning theorists interested in developing an action-centred view of planning to draw on relevant theory on both communication and negotiation. However, this tended not to occur; although in the field of public policy and action generally important theoretical work emerged in the 1980s on the subject of negotiation (e.g. Fisher and Ury, 1981; Raiffa, 1982; Susskind and Cruikshank, 1987), urban planning theorists drew less on this literature and concentrated, instead, on the theory of communication. It is thus significant that Sager (1994) and Innes (1995) termed the new planning theory '*communicative* planning theory' and theory of '*communicative* action' respectively (see also Healey, 1992a). Admittedly, the communicative planning theorists typically adopted a wide, all-embracing view of communication which encompassed debate and argument (e.g. Healey, 1992b; Fischer and Forester, 1993). Even so theorists drew more heavily on communication than on negotiation literature and, in doing so, they concentrated especially on the rather abstract philosophical work developed by the German philosopher and social theorist, Jurgen Habermas. Why was this?

Part of the explanation probably lies in the fact that the planning theorists who developed the new 'communicative' planning theory in the 1980s and 1990s were not *just* interested in the problems of implementation conceived in the narrow sense of 'how to get things done'. Had they been, they might well have focused on the theory and practice of negotiation, for it is primarily through negotiation – through bargaining and reaching agreements with other actors who have the resources to invest in development – that planners can best secure implementation. The communicative planning theorists of the 1980s and 1990s were interested in the problems of action and implementation; they *were* interested in how planning could become more effective in actually 'achieving', and hence 'doing' things. But these planning theorists were also motivated by the ideal of a democratic, participatory style of planning which incorporated all groups who stood to be affected by environmental change, not just those powerful actors who were in a position to carry out – or 'implement' – major development and environmental change. It is this, more than anything, which explains the primary focus on 'communication' in general, and Habermas's theories in particular.

Habermas is a German philosopher and social theorist who has been concerned with developing general theory that provides the grounds for a critique of contemporary capitalist society and, at the same time, provides insights into the preconditions of a more democratic society. It is this latter ideal which has inspired Habermas's theory of 'communicative action'. According to Habermas, if two or more people are to communicate effectively with each other certain conditions have to be met, which Habermas (1979, p. 1) calls 'general presuppositions of communication'. He explains these by suggesting that, when a person A communicates to another person B, A implicitly makes or assumes four 'validity claims'. First, A assumes that what he or she says is *comprehensible* (i.e. its meaning is intelligible) to B. This is obviously a precondition of communication because, if what A says is *in*comprehensible to B, then clearly no communication will pass between A and B. Secondly, A must communicate *something* to B, from which Habermas infers that A must assume (or make the 'validity claim') that the 'something' he or she is saying is factually *true*. Thirdly, for A to communicate to B, it must be A *him or herself* who communicates, from which Habermas infers that A must be *sincere* in communicating to B. In Habermas's terms, the 'validity claim' is that, for genuine communication to take place between two persons, the speaker must not deceive the listener. Fourthly and finally, in order for A to communicate to

B, A must be seeking to come to an understanding *with another person* (B). Accordingly, A must assume that what he or she says is appropriate, or *justified* or *legitimate*, within the context of certain moral norms and conventions shared by both A and B, e.g. about how one person should relate to another (*ibid.*, pp. 2–3).

Set out like this, Habermas's theory may seem purely abstract, and far removed from a practical activity like planning. However, Habermas is clear that his theory has significant practical implications. Habermas advances his theory as one of *practical* reasoning, stressing that often the point of interpersonal communication is to take action. Hence he describes his theory as a theory of communicative *action*.

Habermas's theory enables us to envisage what conditions have to obtain for any contending parties (e.g. in relation to a planning matter) to properly communicate with each other, viz: the linguistic exchanges between them would have to be (to summarise Habermas's four preconditions of communication) *comprehensible*, *true*, *sincere* and *legitimate*. If these four preconditions do not obtain then no genuine communication will take place, and it is in relation to this point that the practical, and political, significance of Habermas's theory becomes clearer. For communication is itself a precondition of real democracy, and hence of any democratic participation in planning, and without genuine communication there cannot be any genuine participation in planning (or any other) decision-making.

Habermas's four conditions of communication therefore furnish us with a normative ideal to aspire to in establishing, and then operating, a participatory process of planning. The ideal is that all communication between participants should be comprehensible, factually true, sincere, and legitimate within a given normative context. Habermas describes this ideal model of communication an 'ideal speech situation'. The practical value of Habermas's ideal model is that it provides us with criteria of communication against which to assess, and if need be criticise, real-life situations of interpersonal communication. His theory enables us to critically examine and assess any real life process of planning which claims to have permitted genuine participation between interested parties.

An illustration of the application of these ideas to planning practice has been provided by Kemp (1980) in his examination of the 'Windscale' local public inquiry held in 1977 into the proposal to build a nuclear reprocessing plant at Windscale in Cumbria. Although this inquiry was advertised as an exercise in public participation in planning, Kemp showed that it left much to be desired when measured against Habermas's criteria of communication. For example, according to Kemp, the final report of the inquiry (which concluded in favour of planning permission being granted for the nuclear plant) contained inconsistencies, omissions and misrepresentations of evidence presented to the inquiry, and so fell foul of Habermas's conditions of factual truth and sincerity. Kemp therefore concludes that the final decision was made on the basis of 'distorted' communication, and was not reached as a result of 'the force of the better argument' (*ibid.*, p. 366).

The leading pioneer of 'communicative' planning theory has been the American John Forester. Forester has drawn extensively on Habermas's theory as a vehicle for evaluating planning practice in terms of the professed ideals of good communication and democratic participation. However, Forester has also exhibited a realistic appreciation of the constraints under which planners work, and in this respect his work connects with our earlier discussion of implementation theory.

In *Planning in the Face of Power* (1989), Forester begins from the twin premisses that 'planning is for people' and, in western democratic societies, planning practice is highly constrained by the 'political realities' of a 'strongly capitalistic society' (*ibid*., p. 3). Forester's aim is to explore what skills planners need to maximise their effectiveness in planning for people 'in the face of power'. Like the implementation theorists, Forester develops the view that, in order to 'get things done', planners have to be effective communicators and negotiators: 'In planning practice, talk and argument matter . . . the day-to-day work of planners is *fundamentally communicative*' (*ibid*., pp. 5, 11, emphasis added). Yet Forester also insists that, in 'getting things done', public sector planning should aspire to the ideal of democratic decision-making over development proposals. Although planners will necessarily be involved in negotiating with powerful developers they should also be active in protecting the interests of all groups in the public, including less powerful or marginalised groups. It is here that Forester draws on Habermas to emphasise the duty of planners to facilitate democratic, participatory planning. In emphasising planners' duty to involve less powerful groups, to expose distorted communication and 'misinformation' and so on, Forester sees planning as a 'communicative' process carrying with it a 'communicative *ethics*' (*ibid*., pp. 22–4, emphasis added):

> by choosing to address or ignore the exercise of political power in the planning process, planners can make that process more democratic or less, more technocratic or less, still more dominated by the established wielders of power or less so. For instance, planners shape not only documents but also participation: who is contacted, who participates in informal design-review meetings, who persuades whom of which options for project development. Planners do so not only by shaping which facts certain citizens may have, but also by shaping the trust and expectations of those citizens. Planners organise cooperation, or acquiescence, in addition to data and sketches. They are often not authoritative problem-solvers, as stereotypical engineers may be, but, instead, they are organisers (or disorganisers) of public attention: selectively shaping attention to options for action, particular costs and benefits, or particular arguments for and against proposals. (*Ibid*., p. 28)

THE POLITICAL ECONOMY, THE MARKET AND IMPLEMENTATION

The concerns of those planning theorists who developed theories about the political economy of planning in the 1970s (see Chapter 6) and the concerns of those who focused on the problems of implementation might seem poles apart.

For theory about the political economy of planning was theory at a very general level and was regarded by some as highly 'abstract'. It was theory which situated and explained the activity of town planning within a general account of the nature of capitalist society and the role of the state within capitalism. Implementation theorists, on the other hand, were concerned with the 'coal face' of planning practice, and how, by understanding and engaging with the real world of action, planners and policy-makers might become more effective actors (or implementers) themselves. Theory about implementation could therefore be seen as highly specific, practical theory; it was about *practical* reasoning. However, both bodies of planning theory accepted that town planning was not an autonomous activity operating in a vacuum, separate from the rest of society. Thus, whether we were concerned with understanding and explaining planning (including its effects) – as the political economists were, or seeking to learn how practising planners could become more effective actors – as the implementation theorists were, it was necessary to view planning as operating within a given social context, in which other actors were operating, and not always with the same objectives as public planning authorities.

The political economic theorists drew attention to the fact that town planning operated within the political economic context of a market system in land and property development, so that what planners did was circumscribed, constrained, even determined by the dictates of the market. Hence Pickvance's (1977, p. 69, in Paris, 1982) thesis that 'the determining factor in urban development is the operation of market forces subject to very little constraint'. It was also a recognition of the part played by other social actors operating 'outside' public sector planning, and especially private sector developers operating within a competitive property market, which was central to an understanding of the problem of implementation in planning. The main reason there was an 'implementation gap' in town planning was that the planning systems of liberal capitalist societies like the UK did not themselves *provide* development, but rather regulated and controlled it. Consequently, the implementation of public sector plans and policies depended in large part on the willingness of private sector developers (and other actors outside the planning system) to come forward and undertake the desired development. Understanding the process of implementation and how to become more effective at implementation therefore required an understanding of the market, and the actors and agencies working within it.

There was, then, this much in common between the very general political economic theory of planning described in Chapter 6, and the apparently more focused and practical concern with implementation which we have examined in this chapter. But there were also important differences which accompanied these two theoretical perspectives, especially in respect of the normative or ideological assumptions which tended to be made by the respective protagonists of these two bodies of planning theory.

Many of the theorists who developed a political economic theory of town planning, and who in so doing emphasised the structural power of the capitalist market system in determining urban development, drew on Marxist social theory to explain the part played by town planning in shaping urban development. These theorists, therefore, were seeking to explain what planning did *as a matter of empirical fact*. They were aiming to develop a social *scientific* theory of planning, a theory which was empirically true and so uncontaminated by the particular values or ideological commitments of the social theorists themselves.[7] However, most of the political economic theorists who relied heavily on Marxist social theory were also committed to Marxism – or some form of socialism – *as a political ideology*. Therefore, the fact that the capitalist market system dictated urban development rather than public sector planning was typically regarded by these theorists as regretable. In this respect, the Marxist inspired political economic theory of planning of the 1970s was presented by its protagonists not only as a dispassionate scientific explanation of the facts, but also as a *critique* of the fact that urban development was so strongly shaped by market capitalism. For ideological Marxists the 'privatised' capitalist market system should be superseded by the *social* ownership and control of the means of production, in land development just as in other areas of production. A stronger system of public sector urban planning would be able to plan development 'positively' without encountering the resistance of private landowners or developers.[8] The idea that town planners should become more effective at implementing plans and policies by co-operating and working 'with' the market system, and hence private sector developers, was therefore far from the minds of these political economic theorists.

Not so, by contrast, with most of those who were responsible for developing a theoretical concern for implementation in planning during the 1970s and after. Here, the message which came across was that, if town planners wished to become more effective at implementing plans and policies, they had to learn how to negotiate and strike bargains with other actors and agencies, including private sector developers – in other words, in a society whose political economic context was a liberal capitalist one, the implementation of public sector plans and policies depended on planners working 'with' the market. And of course this would sometimes involve pragmatically compromising public planning ideals to achieve something which would not otherwise be achieved.

Thus, in the context of a liberal capitalist society, securing the implementation of a public policy sometimes involves compromising substantive planning ideals. For example, a planning authority accepts a less than ideal design for a building in order to secure the development of a site or, in order to reach agreements with developers, planners may have to negotiate deals behind closed doors (we may recall here Bardach's talk of good implementers being 'fixers'). One does not have to be a Marxist to feel uneasy about such pragmatic manoeuvring. Planning theorists like Forester accepted that planners must be prepared to 'get their hands dirty' by negotiating deals with capitalist developers in order to secure some 'planning gains'. But, alongside this, Forester also articulated the Habermasian ideal of making the planning process as democratic as possible by opening up the communicative process of decision-making to all interested parties. In this way, Forester sought to combine the insights of those neo-Marxist political economists who maintained a critical distance from capitalism with the more pragmatic implementation theorists who accepted the need to work with capitalist land developers (and other interest groups) in order to secure at least some planning gains.

It is not only uncompromising Marxists who have been critical of the pragmatism implicit in implementation theory. Other planning theorists have also worried that, in attending to the problems of action in order to 'get things done', there are dangers of planners forgetting about *what* gets done or implemented, or about *how* things get decided. Reade (1987, p. 92) has expressed his unease with situations in which planners negotiate agreements with developers by operating 'as "fixers", "getting things done", and "making things happen" by working behind the scenes'. In such circumstances, Reade (*ibid.*) argues that 'there is obvious scope, if not for outright corruption, then at least for (let us call it) mutual adjustment'. And Reade (*ibid.*) goes on to speculate about what can follow from this:

> In sociological terms, developers and planners will come to develop a shared subculture. They will be likely to develop shared attitudes and values, shared perceptions of what is economically possible and socially desirable, and most significantly, shared beliefs as to what kind of development is in the 'public interest'. They will be constantly working out between themselves, in private, what seems best for the 'community', rather than following standards and objectives imposed upon them as a result of open political debate and formal decisions reached by democratically elected representatives.

The theory and practice of implementation has therefore raised important ethical and political questions about planning. In particular, it has raised the central issue of whether, in the context of a capitalist market system, town planners should engage in practices which involve working pragmatically with the market and compromising certain planning ideals to achieve at least something on the ground. One's view about this depends on one's political ethics and ideology. For those ideologically committed to Marxism, working 'with' capitalist developers is exactly what planners should not do. For according to Marxist ideology, the aim is to replace capitalism, not perpetuate it by striking deals which further the interests of capitalist developers. Political liberals take a more positive view of the market system, and therefore also of a style of planning which works with the market. Given their belief in the virtues of free competitive markets, some strong liberals might argue for a planning system which supports or overtly facilitates market processes. This liberal position came to be espoused openly by some planning theorists in the late 1970s and early 1980s. In the wake of the neo-Marxist political economic perspective on urban development and planning there emerged a neoliberal school of political economy which defended free market capitalism. From this perspective, a planning system which worked with rather than against market forces was not to be apologised for, but rather to be positively welcomed. If, as liberals argued, markets were generally so efficacious, then the question naturally arose as to whether public sector planning was needed *at all*.

This question came to be more than academic in the 1970s, which saw a dynamic resurgence of traditional or 'classical' liberal ideas. In Britain and several other western democracies these liberal ideas came to be expressed by right-wing political parties, and became the central driving force in what came to be described as 'New Right' political ideology. In Britain, this was marked by the election in 1979 of a Conservative government under the leadership of Margaret Thatcher. In its rhetoric this government was determinedly 'promarket' and 'anti-planning', and one of its professed missions was to 'roll back the frontiers of the state'. By the end of the 1970s, therefore, the political economy of market capitalism was brought to bear on the whole idea, and practice, of town planning as it had been established after the Second World War.

NOTES

1. The latter remark here is not just a facetious aside. The history of post-war planning practice was replete with examples of policies and plans which were never implemented, even when they had been approved by planning authorities.
2. See Hill, 1968; McLoughlin, 1969, Chap. 10; Lichfield, Kettle and Whitbread, 1975, Chap. 4.
3. In expressing it like this there is a danger of setting up a false dichotomy between decision-making and action. For 'action' – or implementation – is also an arena of decision-making. The real distinction is therefore between, on the one hand, making decisions about plans and policies without giving any significant consideration to how those plans or policies might be enacted or implemented and, on the other, decision-making about plans and policies in which considerations (and decisions) about action/implementation are included.
4. In this respect, all normative theories of the planning process are (competing) theories of how best to engage in *rational* action. Hence the misnomer of Lindblom's 'disjointed' incrementalism which was discussed in Chapter 4.
5. Recognition of this was of course at the heart of the critique of the 'master' or 'blueprint' view of town planning, which we have met before (see Chapter 3).
6. I say 'usually'. There are two exceptions which we should note. First, policies may sometimes be devised which are *not a response* to a situation, but are a means of realising some ideal or goal we cherish. For example, we may devise a policy to enhance the aesthetic quality of an area which is already attractive. The second concerns cases where policy is devised as a response to a situation, but not as a response to *the actions of others*. For example, policy may be devised as a response to 'acts of nature'; that is, 'actions' (or more properly, 'occurrences') which are not the result of previous actions by other people. Examples would be policies devised in response to natural disasters such as earthquakes, floods, avalanches, etc.
7. There has been considerable debate about whether scientific theories, and especially theories in social science, can be – even in principle – value-free in the way suggested here. I do not go into this philosophical debate here.
8. See, e.g. McKay and Cox (1979, Chap. 2) for an account of this socialist view of positive planning. See also McKay and Cox (1979, Chap. 3) and Boddy (1982, in Paris, 1982, pp. 83–94) for related discussions of the land values issued in relation to the socialist ideal of positive planning.

8

Planning theory after the New Right

INTRODUCTION

In Chapter 6 I described how, in the 1960s, it was openly acknowledged that town planning is a political activity, and hence that the value judgements embodied in plans and planning decisions should be opened up to political debate, including the participation of the public. However, at that time, hardly anyone questioned the desirability of town planning *as such* – that is, hardly anyone questioned the *system* of town planning which had been established after the Second World War, even though this system embodied a particular political view of the role of the state, and state planning, in relation to the capitalist market system. *That* political view, the political position of post-war 'social democracy', was not seriously opened up for debate.

However, in the late 1970s, the fundamental political premise of town planning came to be challenged seriously by an emergent right-wing political movement known as the New Right. This movement was inspired by traditional liberal arguments which favoured the free market and criticised state planning as a 'burden' on private enterprise and efficiency. Although the ideology of the New Right was driven by classical liberal rather than traditional conservative philosophy, it was generally 'conservative' parties which adopted New Right ideas.[1] In Britain, New Right ideas were taken up by Margaret Thatcher after she became leader of the Conservative Party in 1975, and so 'Thatcherism' became the British manifestation of New Right thinking. In 1979, following the largest swing in any general election since 1945 (Sked and Cook, 1993), the Conservative Party under Margaret Thatcher was swept to power, and it was to remain in power for the next 18 years, until the general election of May 1997. Developments in planning theory in the 1980s and 1990s cannot be disassociated from the changes to planning practice brought about by this political shift to the right. In this chapter I describe the impact of Thatcherism on British town planning practice in the 1980s and examine how planning theory has fared 'after the New Right'. Then, I describe the revival of classical liberalism in the 1970s and the liberal critique of post-war social democracy, for it was on this foundation that the New Right built. I then discuss how British town planning practice fared under Thatcherism, before examining the

theoretical responses to the changes which took place in the role (
and planning, through the 1980s. In particular I describe how the d
of 'regime' theory to explain the changing nature of local goverı
'regulation' theory to explain the changing nature of capitalism,
attention (again) to the importance of understanding the politica
context of planning. Both regime and regulation theory are very general theories, but not all planning theorists in the 1980s and 1990s have concerned themselves with such general theorising. Over the last two decades many town planning theorists have turned away from 'grand theory' about planning and turned instead to researching in a more focused way the urban issues and problems which town planning seeks to address. I summarise some of this more specific, 'problem-centred' planning research and theory and the chapter concludes with a brief comment on the state of planning theory now.

SOCIAL DEMOCRACY, THE REVIVAL OF CLASSICAL LIBERALISM AND PLANNING IN QUESTION

Post-war social democracy and the revival of classical liberalism

For thirty years following the Second World War, until the mid-1970s, town planning in Britain (as in many other western non-socialist nations) operated within a political context in which there was a broad consensus between the main political parties of the left and right. This period of post-war politics has been described as a 'social democratic' consensus because, in spite of the differing labels of the major parties ('Conservative' and 'Labour' in Britain), they both subscribed to an ideological position which is best described as 'social democracy'.

Social democracy aimed at a 'middle way' between the extremes of unbridled capitalism (or classical 'free market' liberalism) on the one hand, and Soviet-style state socialism on the other. This social democratic 'middle way' can be seen in the attempt by post-war governments in Britain to create a 'mixed' economy – a mixture, that is, of both the private and the public sector, the market and the state. Thus governments continued to accept private property ownership, private enterprise and free competitive markets as the main institutional mechanisms for running the nation's economy, and to this extent, post-war social democracy accepted liberal capitalism. But post-war social democracy significantly extended the state's role in overseeing and managing market capitalism in order to achieve certain socially desirable goals such as full employment, fair wages, and greater social equality. Part of this greater role for the state included 'nationalising' some strategic industries and services. Part of it involved the state providing, from 'progressive' taxation, a range of basic welfare services such as universal education and health care, social security for the unemployed or those unable to work through sickness or disability, subsidised housing and public transport, and so on (the so-called 'welfare' state).

The social democracy established after the Second World War aimed at a mixture of capitalism and collectivism, of liberalism and socialism. As such, it could be described as 'socially managed capitalism',[2] and the British town and country planning system thus created expressed perfectly the balanced 'mixed economy' approach of social democracy. While the land development industry (i.e. property developers) remained largely in the private sector, the right to develop land was brought under public control by the Town and Country Planning Act 1947 so that henceforth developers were required to apply to the state – in the form of local planning authorities – for planning permission to develop land. In this way, the state was given powers to oversee and regulate the capitalist land market. Here, then, was that characteristic blend of market provision and state regulation which typified post-war social democracy.

For about a quarter of a century, the 'middle-way' of social democracy was so widely accepted that it seemed to have resolved the ideological battles between right and left, between liberalism and socialism. Hence Daniel Bell's (1960) claim that post-war social democracy had brought about the 'end of ideology'. Certainly, some criticisms were made. Nationalised industries were commonly criticised as being bureaucratic and inefficient. Trenchant criticisms were made of some aspects of town planning, such as comprehensive slum clearance and housing redevelopment schemes or the inegalitarian effects of 'urban containment' policy. Yet it was rarely suggested that the state's role in planning should be scaled down or removed altogether, and criticism was usually followed with proposals for alternative planning policies not the abandonment of planning altogether.[3] Thus Young and Willmott's critique of housing redevelopment in east London concluded with the suggestion that planners be more sensitive to the communities subject to rehousing schemes by, for example, rehousing communities *en bloc* so that they were not fragmented in the redevelopment process. Similarly, Peter Hall and his colleagues followed their criticisms of the inegalitarian effects of urban containment policy with suggestions for alternative policies (Hall *et al.*, 1973, Chaps. 12 and 13). So although its operations were inevitably imperfect, the interventionist role of the post-war social democratic state was generally accepted, and criticism was directed at improving its operations rather than doing away with it altogether.[4]

However, although it seemed to some (such as Daniel Bell) that social democracy had seen off its ideological rivals, and notably classical 'free market' liberalism, even in the heyday of post-war social democracy some political theorists and commentators continued to argue for a return to a 'free market' economy. One of these was the economist and political theorist, Friedrich Hayek, who had been a critic of central state planning since the 1940s (Hayek, 1944). According to Hayek large-scale centralised planning requires a degree of knowledge and information which even the most enlightened officials cannot collect and use intelligently to make sound decisions. Under a market economy, information and decision-making are 'decentralised' amongst the numerous firms and organisations operating within the market and responding to its signals. In addition to this 'technical' reason for preferring free markets to a planned economy, Hayek also opposed centralised planning on political grounds, contending that 'planning leads to dictatorship' (*ibid.*, p. 52). An influential figure in British politics was the Conservative Enoch Powell who despite his traditional conservative views about maintaining cultural identity by restricting immigration, throughout the 1960s advocated a vigorous brand of free market liberalism (see, e.g. Powell, 1969).

It was largely economic concerns which fuelled the revival of classical liberal ideas in the 1970s when the post-war social democratic consensus came under increasing strain under the pressures of inflation, sluggish economic growth (especially in Britain), and the high levels of public expenditure needed to fund the welfare state (O'Connor, 1973). Liberal theorists attributed the economic malaise of social democracy to excessive state intervention which led to inefficient 'bureaucratic' decision-making and stifled private enterprise, competitiveness and efficiency. The high levels of taxation to finance public spending on the welfare state were also said to be a burden on the 'productive' private sector. The American economist Milton Friedman (1962) argued for a return to classical liberal principles of fostering free competitive markets. Liberals also contended that free markets were more compatible with the liberal ideals of individual freedom and responsibility. The American philosopher Robert Nozick (1974) argued that a proper respect for individuals' rights to liberty required some form of classical 'Lockean' liberalism, with only a 'minimal' state. These were the sorts of ideas that were revived and promulgated by the New Right, and the resurgence of classical liberalism became influential throughout western Europe and North America in the 1980s. In the United States, for example, 'Reaganism' was the counter-part to 'Thatcherism' in Britain. This revival of free market liberalism was subsequently reinforced by the collapse of the Soviet system of socialism throughout eastern Europe and the Soviet Union itself. Although many socialists insisted that the kind of socialism established by the USSR was not 'real' socialism, the fact remained that the wholescale discrediting of 'actually existing' socialism put most socialists on the defensive. Apart from anything else, it was widely acknowledged that, in addition to its oppressive authoritarianism, Soviet socialism was brought down by the manifest failure of an economic system in which the means of production (including the means of 'producing' land development) had been totally under 'social' (i.e. state) ownership and control. Even socialists unsympathetic to Soviet socialism had continued to hold that a genuinely socialist society required the social ownership or control of the means of production, and so the discrediting of this left many socialists at sea. It is no wonder, then, that the 1980s saw many western socialists seeking to re-establish the credibility of socialism by advancing versions of 'market' socialism (e.g. Nove, 1983; Le Grand and Estrin, 1989; Miller, 1989). However the word 'market' here made it clear that this was a defensive move – an acknowledgement of some concession to, and thus victory for, liberal market economics.

The rise of the neo-liberal New Right in western democracies, together with the collapse of socialism in the Soviet Union and its satellites, therefore reinforced the groundswell of opinion that liberal, competitive markets were a more effective means of organising production and consumption than socialist planning, and, by extension, the public sector planning of social democracy too. Such was the confidence in the 'triumph' of liberalism that an American political theorist, Francis Fukuyama (1989, p. 4), announced that liberal democracy marked 'the end point of mankind's ideological evolution' – the 'end of history', as he dramatically put it. This sounded like a restatement of Daniel Bell's 'end of ideology' thesis, and indeed it was. Only, by 1989, it was a much more severe free market version of liberalism that was being proclaimed as triumphant rather than Bell's social democracy of the 1960s.

Market liberalism and town planning

The revived classical liberal thinking was also applied to town planning. In fact, back in 1960, in outlining his liberal theory of political economy, Hayek (1960, pp. 341–2) had written about town planning:

> the market has, on the whole, guided the evolution of cities more successfully, though imperfectly, than is commonly realized and . . . most of the proposals to improve upon this . . . by superimposing a system of central direction, show little awareness of what such a system would have to accomplish, even to equal the market in effectiveness'.

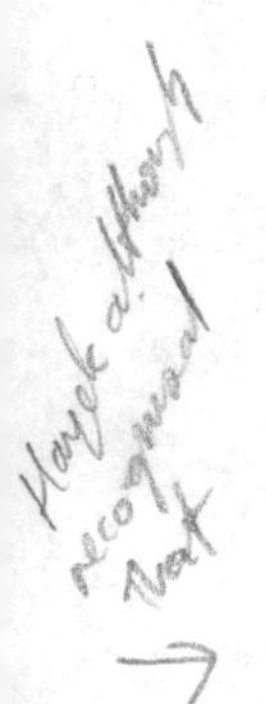

Although this was Hayek's general position, he did consider that the state should play a significant role in planning or at least in intervening in the urban land market. For where people and activities reside in close proximity Hayek (*ibid*., p. 341) argued that the free play of market forces and private enterprise 'do not . . . provide an immediate answer to the complex problems which city life raises'. This is because activities in close proximity give rise to numerous positive and negative 'neighbourhood effects' or 'externalities' which are experienced by their recipients as 'windfall' gains or losses. Because these benefits and harms are not 'earned' by the people who experience them, there is, according to Hayek, a *prima facie* case for a system of public intervention to neutralise these neighbourhood effects, just as – according to liberal theory – the state should intervene to protect people's rights from uncalled-for assault, theft and so on. Accordingly, Hayek proposed a system of financial intervention which would involve taxing the benefits (such as rising property prices) accruing from positive neighbourhood effects, and compensating those who suffered negative neighbourhood effects. Interestingly, this proposal was virtually equivalent to the financial provisions for taxing 'betterment' and compensating 'worsenment' which the post-war Labour government attached to the Town and Country Planning Act 1947.

Notwithstanding this, Hayek stuck to his liberal view that urban development should be governed by market forces, not planned by the state. This liberal position was revived in the 1980s by a number of planning theorists. Robert Jones (1982) put forward a more extreme case than Hayek for urban development to be left to the market, citing with approval the example of Houston, Texas, which had no local planning system and yet which, according to Jones, suffered none of the problems planning was supposed to solv prevent. On this basis, Jones advocated the virtual dismantling of the Britis town and country planning system, leaving only legal controls that operated through the law of nuisance and private covenants to control unwanted developments in residential estates.

Anthony Sorenson and Richard Day offered a more moderate position than that put forward by Jones, but still they argued for an essentially market-led system of urban development (Sorenson and Day, 1981; Sorenson, 1982; 1983). They suggested that most planners had hitherto assumed a socialist – or at least social democratic – view of planning, but that liberal (or as they termed it 'libertarian') political philosophy offered a valid alternative to this:

> libertarian [or liberal] philosophy provides a valid alternative (and indeed more practical) conception of how western social formations are, or ought to be, structured in contrast to the more collectivist (and perhaps socialist) views traditionally espoused by planners . . . An appreciation of libertarian philosophy . . . can provide a valuable corrective to planners' tendency to regard more planning and control as better. (Sorenson and Day, 1981, p. 390)

Like the neo-Marxist political economists before them, Sorenson and Day acknowledged that town planning had to be seen within its political economic context, and that in western liberal societies this context is a capitalist market economy. But unlike Marxists, Sorenson and Day (1981, p. 391) follow Hayek in celebrating, rather than lamenting, this fact:

> Markets . . . are essential for freedom because they permit the individual maximum freedom of choice . . . Individual freedom per se is not the only benefit claimed for a free society. Freedom encourages greater creativity, which in turn makes for more interesting and varied life-styles, more invention and innovation, and greater flexibility in the face of changing circumstances. Market forces tend to be much better than economic planners in increasing society's volume and quality of output and in rewarding merit and effort. In short it can be argued that social formations which emphasise individual liberty are generally more creative, productive, dynamic, responsive to human needs, and flexible than those which do not.

In spite of their general endorsement of markets, Sorenson and Day seem to accept the need for some basic system of town planning by the state, in the form of 'notional land use zoning'. But this planning should generally support, rather than act as a countervailing force against, market-led development. As they put it (*ibid*., p. 393): 'the planners' function is either to correct malfunctions in the market place or to facilitate market processes. Planning activities are therefore mainly economic in kind and require a sound understanding of price mechanisms and nature of entrepreneurship.'[4]

How influential these theoretical ideas were on the first Thatcher government which came to power in 1979 is difficult to assess. Certainly the agenda of that government was shaped by the revival of classical liberal ideas

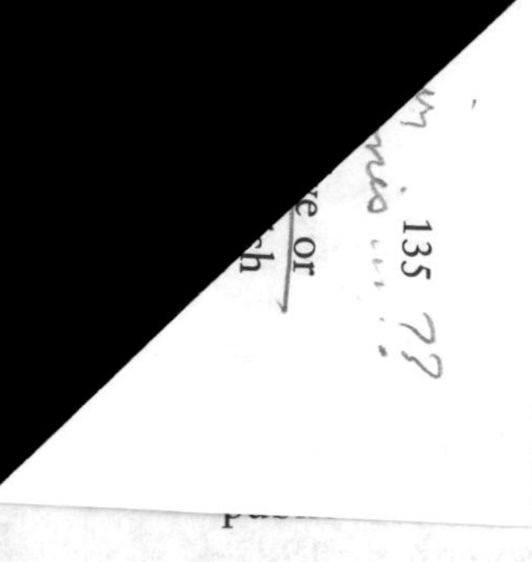

section, as was evident in its rhetoric of 'rolling back the ;tate'. As part of the social democratic state, town planning nplicated in this; indeed, state planners were singled out for Some even wondered whether, after a few years, there might ınning system left at all, never mind any town planning theory. :refore presented the biggest challenge to the whole idea of vn planning since the end of the Second World War.

BRITISH TOWN PLANNING UNDER THATCHERISM

Four months after the May 1979 election, the first Secretary of State for the Environment appointed by Margaret Thatcher – Michael Heseltine – was invited by the Royal Town Planning Institute (RTPI) to address its summer school. Heseltine used the occasion to outline his view of the role of the town and country planning system, and this was to set the framework of the Conservative government's strategy towards planning throughout the 1980s.

Prior to his RTPI address, and as part of his own commitment to the government's neoliberal strategy of 'rolling back the frontiers of the state', Heseltine had announced plans for scrapping or relaxing over 300 controls exercised by central government. This had fuelled rumours that the Conservative government might be about to dismantle planning controls too. In his address Heseltine straightway countered these rumours, saying that he had 'no intention of wrecking the planning system developed in the last 40 years or so in this country'. Rather, his intention was to reform and improve the system by bringing it 'up to date' (Heseltine, 1979, p. 25). Heseltine meant two things by this.

First, as far as planning procedures were concerned, Heseltine held that these were unnecessarily slow and cumbersome, largely because the operation of the system was encumbered by attention to detail at the expense of basic principles. He lamented the slow progress of structure plan preparation, where, over the five years since 1974, only 57 of a possible 72 structure plans had been submitted to the government for approval, and of these only 27 had met with approval. Given that structure plans were designed to lay down the planning strategies for localities 'in broad-brush terms', Heseltine could not understand why 'it should take a planning authority so long to paint these broad brush strokes' (*ibid.*, p. 26). Nor could he see why it had taken 'what is now my Department so long to give approval' to those plans submitted (*ibid.*). Similarly with the control of development, Heseltine argued that the system had become 'ponderous and detailed'.

Heseltine therefore declared his intention of 'streamlining' planning procedures to make them more 'efficient, responsive and speedy' (*ibid.*, p. 25). The key to this was to distinguish major from minor issues, in both the making of plans and the control of development, and then adopt different procedures for each, e.g. by ensuring that development plans concentrate only on the main issues; by delegating applications for minor developments to officers, rather than having them go through a full planning committee; by enlarging the

category of development which, by virtue of its relative insignif
not require planning permission; etc.

Secondly, as far as the substance of planning decisions wa
Heseltine held that planning authorities should take a much mo
view of applications for planning permission; in other words,
should take a 'positive' view of market-led development. He was
planning system which did 'not act as a drag on the necessary processes of development', and he famously added that 'thousands of jobs every night are locked away in the filing trays of planning departments' because of delays in issuing approvals for development (*ibid*., p. 27). Heseltine saw the planning system as contributing to the government's general economic strategy of regenerating the British economy by 'lifting the burden' of state restrictions. As he said in the conclusion to his speech (*ibid*., p. 29):

> In recent decades we have all seen a steady decline in the state of the British economy . . . There are many reasons why we have apparently inexorably slid down the economic league table. Not for one moment am I suggesting that planning or the planning system is the main factor in this. But it is one contributory factor among many.

During the 1980s, three successive Thatcher governments pursued a strategy towards planning which was largely in line with the Heseltine agenda. Thus, in both its legislative changes and circular advice to local planning authorities, the governments sought to streamline planning procedures and to bring about an approach to planning control which, outside conservation areas, was largely supportive of market-led development proposals. But how significant were these changes? In particular, did Thatcherism significantly *alter* town planning in Britain?

A decade of Thatcherism did not, in fact, alter significantly the *system* of planning control which had been put in place in 1947, although the government did bring in new planning legislation and introduced a number of innovations, one of which involved by-passing the established planning system by setting up urban development corporations (UDCs) to regenerate some inner areas of Britain's major cities (see, e.g. Thornley, 1991, Chap. 8). The government also legislated to simplify planning regulations by, for example, changing the Use Classes Order which specifies what constitutes a change in land use (and hence what counts as 'development'); enlarging the category of permitted development not requiring planning permission; and – perhaps most significant – introducing 'enterprise zones' (EZs) which suspended many of the normal planning regulations in the areas to which they applied (*ibid*., Chap. 9). Some of these innovations, such as UDCs and EZs, however, were designed to operate for only a temporary period. Others, such as the changes to the Use Classes Order, represented relatively minor tinkerings to the established system. Moreover, although planning control was weakened in the early years of Thatcherism, the later 1980s saw something of a backtracking from the 'hardline' free market strategy, even before Margaret Thatcher fell from office in 1990. This was partly brought about by a sudden upsurge of environmental ncern, arising mainly from evidence about global warming and ozone depletion, and the feeling that the state – including statutory town and country planning – had an important role to play in environmental protection. One outcome of this was the Town and Country Planning Act 1990. Section 54A of which reinstated local authority development plans as the primary basis for guiding and considering development proposals. In spite of the initial rhetoric, then, Thatcherism did not fundamentally restructure the British planning system.[5]

However, although the planning *system* remained intact, it can be argued that Thatcherism, had a significant *ideological* impact. In his 1979 address to the RTPI, Heseltine had said that his intention was not so much to alter the planning system but rather the way the system worked, by 'streamlining' its operations and making it more supportive of the market. In line with this objective, alongside its (arguably minor) legislative alterations to the actual planning system, much of the government's energy went into the production of circulars (subsequently Planning Policy Guidance Notes, or PPGs) advising local planning authorities how to operate the system. A whole clutch of circulars were issued in the 1980s by the Department of the Environment on such issues as land for housing, green belts, the use and wording of planning conditions, aesthetic control and so on. The overall thrust of these circulars was that planning authorities should take a 'positive' view of applications for development, and hence be more supportive of the market system which generated these proposals.

An early circular on development control – Circular 22/80 (DoE, 1980) – was especially important for in places it echoed Heseltine's speech to the RTPI. Thus paragraph 3 (*ibid*., emphasis added) said that:

> The planning system should play a helpful part in rebuilding the economy. Development control must *avoid placing unjustified obstacles in the way of any development* especially if it is for industry, commerce, housing or any other purpose relevant to the economic regeneration of the country . . . Local planning authorities are asked therefore to pay greater regard to time and efficiency; to adopt a more positive attitude to planning applications; to facilitate development; and *always to grant planning permission* . . . unless there are sound and clear-cut reasons for refusal.

Admittedly, government circulars are only advisory and so lack statutory force. In theory local authorities could still exercise discretion in considering whether or not to follow the advice but in practice they were effectively obliged to heed the advice given. For if an application for planning permission was refused, the applicant could appeal to the Secretary of State for the Environment and, as the very last sentence of Circular 22/80 made clear (*ibid*., para. 21): 'When such appeals are made the Secretaries of State will be very much guided by the policy set out in this circular.' Indeed they were. Throughout the 1980s, major land developers became increasingly bullish in making appeals against refusals of planning permission, and increasingly these appeals were

upheld by the government. The circular thus became a key instrument of central government policy and the main means by which the government realised Heseltine's promise to retain the planning system but to ensure that it functioned to facilitate market-led development.

It is in this way that Thatcherism, whilst only tinkering with the legal instruments of planning control, nevertheless had a significant impact on town and country planning in the 1980s. Arguably, Thatcherism altered the whole culture of planning so that, by the end of the 1980s, planners increasingly saw themselves as partners working 'with' the market and private sector developers. They had little choice to do otherwise, whatever their political views about the role of town planning, for the political context of town planning had changed.

How, then, did theorists of planning respond to this changed political world of planning practice?

THE POLITICAL ECONOMY OF TOWN PLANNING REVISITED: REGIME AND REGULATION THEORY

With the resurgence of neo-liberalism under the New Right, coupled with the collapse of socialism in the Soviet Union and its satellites, by the end of the 1980s it became abundantly clear that town planning in western democracies would have to continue to operate within a capitalist market economy. This recognition was not new. As we saw in Chapter 6, this was central to the Marxist inspired political-economic theories of planning which developed in the 1970s, and in that light the idea that New Right Thatcherism fundamentally altered town planning practice can be dismissed as an exaggeration. But if a decade of neo-liberalism did not fundamentally restructure the *powers* of (public sector) town planning, the explicit requirement on planning authorities to adopt a 'positive' view of market-led development proposals resulted in some significant changes to the way planning was *approached and practised*.

The changes that occurred under Thatcherism can be better understood if we look beyond the confines of statutory town planning practice to some of the wider changes which took place – first, in local government practices and, secondly, in capitalism itself. Changing conceptions of town planning over the last two decades have been in many ways a reflection of these wider changes in local governance and the political economy of capitalism. Two general theories which seek to explain these wider changes are regime theory and regulation theory respectively.

Entrepreneurialism in local government, and regime theory

For the thirty-years period of the post-war social democratic consensus, a significant proportion of the state's public spending went to local government. However, during the crisis years of the mid-1970s (when a Labour government was still in power) public spending was successively cut back, and this obviously hit local government. Whereas local governments – especially large city authorities – had once seen themselves as the main providers of local services, increasingly they had to find ways of attracting the private sector to assist them. Further, in the name of efficiency, central government increasingly required local government either to contract out some of its services to the private sector or to compete with the private sector (through compulsory competitive tendering or 'CCT') for the right to continue providing them. The Thatcher governments' cajoling of town planners to take a positive attitude to market-led development was thus part of a more general strategy of reducing the role of the state (the public sector) and putting more emphasis on private enterprise and competition.

The economic context in which local government operated also changed in another way. As a result of major economic restructuring, many cities in the 1980s lost some of their industries, creating severe unemployment. Local governments turned their attention to what they might do to regenerate their local economies, though with cutbacks in public spending they had to look to the private sector to help them. The net result was a shift in the style of urban governance from what David Harvey (1989) has called the 'managerial' approach of the 1960s to the 'entrepreneurial' approach in the 1980s.

One expression of this entrepreneurial approach has been the forming of 'partnerships' with private sector agencies and developers to accomplish things local government would be unlikely to achieve if it worked on its own. Here the lessons of implementation theory, which emphasised the need for inter-agency networking and agreement, were influential. But so too was the clear acknowledgement that local governments were no longer autonomous agents, able to achieve their objectives by acting alone. In order to maximise their effectiveness, therefore, local governments had to make contacts and negotiate agreements with other non-governmental institutions and agencies. Local authorities thus had sometimes to compromise their aims, but this was (and is) inevitable when several partners with differing objectives come together to act collectively.[6]

'Regime theory' seeks to explain this changing approach to governance. Initially developed in the USA by Clarence Stone (1989; 1993), its starting point was to question the assumption that the power to bring about desired outcomes involves exercising 'social control' (or 'power over' others) from the centre – from government. Stone advances a 'social production' model in which power is seen as the capacity to achieve certain ends ('power to' rather than 'power over'): 'The power struggle [between rulers and challengers] concerns, not control and resistance, but gaining and fusing a capacity to act – *power to*, not *power over*' (Stone, 1989, p. 227).

If the power to achieve certain ends depends on the possession of the capacity to act, and if no government has a monopoly of power (or a monopoly of the capacities to act), then it follows that, to achieve their objectives, all governments must enlist the support and co-operation of other agents, including non-governmental actors. This 'assembling' of power (or the capacity to act) necessarily involves establishing coalitions or partnerships with other agents. Again, Stone contrasts this way of looking at the exercise of government power to that which sees government acting independent of others, and exercising power over others by 'command and control':

> 'Governing', as used in governing coalition . . . does not mean rule in command and control fashion. Governance through informal arrangements is about how some forms of coordination of effort prevail over others. It is about mobilizing efforts to cope and adapt; it is not about absolute control. Informal arrangements are a way of bolstering (and guiding) the formal capacity to act. (*ibid*., pp. 5–6)

Stone calls these governing coalitions 'regimes'. Regimes are 'the informal arrangements by which public bodies and private interests function together in order to be able to make and carry out governing decisions' (*ibid*., p. 6). Or, as Stoker and Mossberger (1994, p. 197) put it: 'Urban regime theory starts with the assumption that the effectiveness of local government depends greatly on the cooperation of nongovernmental actors and on the combination of state capacity with nongovernmental resources. As the task of governments becomes more complex the cooperation of various nongovernmental actors is required.'

According to DiGaetano and Klemanski (1993), one of the reasons regime theory developed first in the USA was the fact that US local government power has traditionally been more limited than in the UK. However, under the Thatcher governments local government power in Britain was progressively weakened by financial cuts and a narrowing of local government responsibilities (*ibid*., p. 65). In this hostile climate local authorities came to see that their capacity to achieve their objectives depended increasingly on forming partnerships with non-governmental agents, and in any case, they were deliberately encouraged to do this by central government as part of its neo-liberal agenda of enhancing the role of the market and the private sector. These changes have meant that regime theory has now become more applicable to the UK.

DiGaetano and Klemanski cite the city of Birmingham as an example. When faced, in the 1980s, with its worst economic crisis in the twentieth century, Birmingham City Council decided to pursue an aggressive strategy to attract business to Birmingham. A central part of this was the involvement of local business as 'a fully participating partner in the city's governing coalition' (*ibid*., p. 70). A number of public–private partnerships were established to carry out various economic development projects. Amongst these was the partnership between the city council, Aston University and Lloyds Bank to develop Aston Science Park as a high-technology business and research park. Another involved a partnership between the city council, the Birmingham Chamber of Commerce and Industry, and five private developers in the Birmingham Heartlands project to regenerate a large area of east Birmingham.

All this could give the impression that regime theory attributes autonomy to local governing coalitions and thereby underestimates the extent to which government, whether acting alone or in coalition with others, is constrained by larger economic forces and powers. However, this would be to misunderstand the theory. Regime theory begins from the premise that, in liberal societies, many of the most significant decisions affecting people's lives are made outside government by firms operating within the capitalist market system. Hence regime theory situates the capacities of local government within the wider political economy. Stone (1993, p. 2) emphasises this in contrasting regime theory to traditional pluralist accounts of politics: 'Urban regime theory takes as given a liberal political economy . . . [and] . . . the economy of a liberal order is guided mainly, but not exclusively, by privately controlled investment decisions.'

Indeed, as we have seen, the whole reason why local governments have had to learn to work with other, non-governmental private sector agencies, is because they are operating within the context of a liberal capitalist political economy, and therefore do not possess absolute power to accomplish things by themselves. However, Stone warns against a kind of economic determinism which would suggest that the actions of local governments are completely determined by the prevailing political economic context. Stone (*ibid.*) insists that politics *matters* – that, by acting creatively in consort with others, local governments can achieve ends they would otherwise not achieve:

> Regime analysis posits a . . . complex process of governance. Specifically, it recognises the enormous political importance of privately controlled investment, but does so without going so far as to embrace a position of economic determinism. In assuming that political economy is about the relationship between politics and economics, not the subordination of politics to economics, regime analysts explore the middle ground between, on the one side, pluralists with their assumption that the economy is just one of several discrete spheres of activity and, on the other side, structuralists who see the mode of production as pervading and dominating all other spheres of activity, including politics.

Even within a capitalist market economy, however, economic conditions can vary, and this can give rise to varying local government strategies. In other words, different kinds of regimes can emerge. Stone (*ibid.*, pp. 18–22) distinguishes between four kinds of regimes: 'maintenance regimes' – which concern themselves primarily with maintaining an inherited state of affairs; 'development regimes' – which seek positively to intervene to promote economic growth or counter decline; 'middle class progressive regimes' – which tend to focus on such issues as environmental protection, historic preservation, affordable housing, the quality of urban design, etc.; and 'regimes devoted to lower-class opportunity expansion' through improvements to education and job training, access to transport and home ownership, etc.[7]

Stone's analysis suggests that different town planning regimes may arise in different political and economic circumstances and this was emphasised by Brindley, Rydin and Stoker (1989) in their analysis of British planning during the Thatcher years. They reiterate the point that, under Thatcherism, there was no major reform of the inherited town planning system, so that Thatcherism in no way abolished planning. Rather, the Thatcher governments called upon planning authorities to use the prevailing planning powers to facilitate market-led development. However, as Brindley, Rydin and Stoker also point out, what

Perceived nature of urban problems	Attitude to market process	
	Market-critical: redressing imbalances and inequalities created by the market	Market-led: correcting inefficiencies while supporting market processes
Buoyant area: minor problems and buoyant market	regulative planning	trend planning
Marginal area: pockets of urban problems and potential market interest	popular planning	leverage planning
Derelict area: comprehensive urban problems and depressed market	public-investment planning	private-management planning

Figure 8.1 A typology of different planning styles in the 1980s
Source: From Brindley, Rydin and Stoker, 1989, Table 2.1

emerged through the 1980s was not just one type of market-supportive planning, but different styles – or 'regimes' – of planning in different places. Their typology of these regimes is reproduced in Figure 8.1. This makes clear that different planning regimes reflected different political positions in different local authorities (different attitudes to market processes) and different economic circumstances (ranging from buoyant to 'derelict' areas).

Amongst the different styles of planning they identified are some that were opposed to the market-led strategy of Thatcherism and which sought instead to adopt an approach in which the public sector played the leading role. The clearest example of this is the public-investment planning adopted by Glasgow City Council in its Glasgow Eastern Area Renewal (GEAR) scheme. However, we should not conclude from this that local planning regimes in the 1980s depended entirely on the ideological complexion of the local council, or that the market-supportive approach to planning favoured by Thatcherism only emerged in areas where the relevant local authority were in agreement with this.

Although a number of local authorities did take a stand against the Thatcherite approach and sought to foster forms of 'local socialism' (Boddy and Fudge, 1984; Gyford, 1985), the government's circular advice backed up by the adjudication of planning appeals brought pressure to bear on local authorities to adhere to their neoliberal strategy. (The strategies pursued by the Greater London Council (GLC) and Sheffield City Council are examples of councils which attempted to pursue local socialism, as are some of the instances of local 'popular' planning described by Brindley, Rydin and Stoker

(1989, Chap. 5). See also Montgomery and Thornley (1990, Chaps. 12–15).) When faced with opposition to this strategy by the 'socialist' metropolitan authorities, such as the GLC, the Thatcher administration abolished these local regimes altogether!

Regime theory acknowledges that wider economic forces remain as central determinants of what local authorities can do. The 'political economic' fact remains that the investment decisions made by non-governmental organisations and firms continue to play a crucial role in shaping the fortunes of localities (see Cooke, 1989). As Brindley, Rydin and Stoker (1989, p. 128) point out in relation to the GEAR scheme, the impact of Glasgow's public sector-led strategy was 'diluted by a further decline in the local economy' such that the GEAR area's rate of decline continued to be more rapid than that for the city as a whole. All local authorities, whatever their political persuasion, have to attract private sector development, and it is this shift to entrepreneurialism that characterises urban governance and planning since the 1980s.

Changes in capitalism and regulation theory

These changes in economic life were further reinforced by changes within capitalism itself. Some theorists have argued that, since the mid-1970s, capitalism generally has been moving towards a new 'mode of regulation' – sometimes described as a move from a 'Fordist' to a 'post-Fordist' mode of regulation (Amin, 1994). This move has involved firms in making 'a ferocious effort to reduce costs and raise productivity' (Fainstein, 1995, p. 37) in response to the perceived 'crisis of profitability' of the mid-1970s, itself attributed to the high wage and tax costs borne by business during the years of post-war social democracy and Keynsian economic policy. A major feature of this has been the emergence of powerful multinational corporations, creating a 'global' rather than a national economy. The drive towards ever-increasing efficiency of production (as measured by the ratio of output to costs) has led these corporations to disaggregate different aspects of industrial production and locate them in places where productivity can be maximised, such as in Third World countries where labour costs are cheaper. Advanced information technology has further contributed to a shift from traditional industrial activity to economies dominated by the provision of services, including financial services (hence the talk of the 'postindustrial' economy).

In Britain the effect of these changes has been a decline of manufacturing industry in the old industrial heartlands and a corresponding expansion of the service sector and 'high-technology' industries in the south (though service industries have also been subject to 'rationalisation' and 'down-sizing', generating unemployment amongst service and professional workers). In these fiercely competitive circumstances, city authorities have had little choice other than to adopt a competitive, entrepreneurial approach so that – and however uncongenial it might be to speak of 'economic determinism' – the political economy of post-Fordism has become a greater determinant over local policy-making and planning. As Fainstein (*ibid.*, p. 38) puts it: 'Cities, like private corporations, are increasingly in the business of making deals. But the kinds of deals public officials can make are limited to what conforms to business strategies.'

However, even in these changed economic circumstances, as Keith Bassett (1996, p. 553) points out, there are both optimistic and pessimistic scenarios. On the optimisitic side the emerging post-Fordist world may provide opportunities for cities to compete for new economic activity and development, thereby emphasising the importance of local politics in shaping their fortunes. The selling of Glasgow as a 'city of culture' or attracting new economic development to Birmingham may be indicators of these possibilities. The pessimistic scenario suggests that, as a result of competition between cities, whilst some may succeed in becoming privileged localities for economic and cultural development, the majority will be losers, and what gains are made in attracting investment will remain precarious and unstable, dependent on the investment decisions of multinational companies with their headquarters elsewhere.

As Bassett (1996, p. 553) says: 'No clear and general answer to the question as to which scenario is most likely can yet be given. We need many more detailed case studies in different contexts to see what kind of room for manoeuvre different cities may have.' However, one thing is clear: in a world where local planning authorities have become increasingly dependent on non-governmental agents to realise their goals, the activity of planning is likely to become increasingly an activity of networking, bargaining and negotiation – of 'doing deals'. It is not surprising, therefore, that implementation theory, and particularly that branch of it which sees planning as a communicative and negotiative activity (see Chapter 7), has continued to occupy a dominant position in contemporary town planning theory.

PROBLEM-CENTRED PLANNING THEORY

To the extent that governments, and town planning authorities in particular, do possess some 'room for manoevre', what should they try to achieve? And, to the extent that town planning is viewed as a negotiative activity, what should town planners negotiate *about*? These are normative questions, for they are questions about the aims, and hence the values of planning.

Most of this book has described the development of town planning theory since 1945 in terms of the development of differing conceptions of town planning as an activity. However, this very general, abstract theorising about the nature of planning (and specifically procedural planning theory) was criticised by some planning theorists who took the view that planning theory should be grounded in the study of the 'substantive' issues with which planning deals. From the late 1970s onwards, therefore, many planning theorists turned away from 'grand theorising' about planning, and sought instead to research and develop theory which focused more sharply on the issues and problems which town planning (and public urban policy generally) seeks to address. As well as being more specific and 'concrete', this theoretical work has been driven by certain value judgements about what the main problems (or aims) of public sector planning are. Although the years of Thatcherism were generally hostile to town planning practice, and although, too, many planning schools were closed down by the government in the early 1980s, the period since the late 1970s has been a fertile one for the more focused kind of planning research and theory just described.

No doubt, one of the things which prompted this more specific, problem-centred research was the emergence of the New Right in politics, and its advocacy of a vigorous free market strategy as the vehicle for addressing economic and social problems. As the New Right pursued its agenda through the 1980s, research interest was naturally drawn to the question of whether its market strategy was proving any more effective than public sector planning in solving the problems which had been the traditional concern of town planning. Even if one had subscribed to the neoliberal doctrines of the New Right one would have been naive to suppose that all the problems which planning had sought to address would go away under the free market. Problems of economic decay and depression in some areas, the relative poverty and lack of opportunities experienced by some social groups, and the degradation of environments both ecologically and aesthetically continued to persist. This drew some urban and planning theorists to reflect on these problems and the roles of the state and the market in addressing them.

What follows is a broad overview of the main issues that problem-centred research sought to address. This account is structured in terms of five main problem areas:

1. The continued economic decline of some (particularly inner) urban areas and the development of theory concerned with urban economic regeneration.
2. The persistence of social divisions and inequalities in society and the theoretical concern over planning for disadvantaged groups and equal opportunities.
3. The discovery of life-threatening ecological changes to the natural environment and planning for environmentally 'sustainable' development.
4. The revived concern for the aesthetic quality of urban environments and renewed theoretical work on urban design.
5. A continuing concern with the degree to which land development and planning is open to local democratic control and hence a continuing theoretical concern with participatory planning.

Urban economic decline and regeneration

In 1977, the Labour government published a white paper, *Policy for the Inner Cities* (DOE, 1977), which focused on the persistent problems of the inner areas of Britain's major industrial cities. This proposed addressing inner-city problems in terms of legislation and resources (e.g. to support firms in inner-city areas) and new institutional arrangements (e.g. an enhanced role for the Department of the Environment; new partnership schemes between central government and the relevant local authorities). Above all, it acknowledged that the key to regenerating depressed inner-city areas lay in *economic* renewal and development (Stewart, 1987, p. 133).

Within two years of this white paper, the government was replaced by Margaret Thatcher's first Conservative government committed to a neo-liberal economic policy with a reduced role for the state. However, the new government maintained, with some 'streamlining', the general arrangements of the strategy instituted by Labour (such as the partnership schemes with local authorities). The Conservative government, though, injected into the policy a whole set of new initiatives designed to make economically depressed inner-city areas attractive for private sector, and therefore market-led, development and regeneration (e.g. Urban Development Corporations, Enterprise Zones, City Action Teams and Task Forces, City Challenge and, most recently, the Single Regeneration Budget). With these initiatives the government hoped, in time, to 'reactivate the free market' as the main agent of inner-city regeneration. (See e.g. Stewart, 1987; Oatley *et al.*, 1995; Lawless, 1996, for assessments of the effectiveness of these policies.)

In spite of the many urban policy initiatives of the 1980s and 1990s, the massive social problems of the inner cities have persisted and, in real terms, the amount of resources directed at inner-city regeneration has declined since the late 1970s (Oatley, 1995, p. 263). The government's urban policy throughout this period also attracted criticism, even from the Church of England (Archbishop of Canterbury's Commission, 1985). However, apologists for Conservative government policy argued that the ongoing plight of inner cities cannot be entirely attributed to a misconceived urban policy in the circumstances a general economic recession in the 1990s, with ongoing high unemployment and social fragmentation.

Whatever the final verdict, the problems of the inner cities have naturally continued to attract the attention of urban theorists and researchers. A considerable literature has been generated on this topic, some of it evaluating the effectiveness of government urban policy in general terms (e.g. Oatley *et al.*, 1995), some the effectiveness of particular policy initiatives such as urban development corporations (e.g. Imrie and Thomas, 1993), and some of it speculating on how urban regeneration policy might be improved in the future, including what part (if any) public policy and planning might play in this (e.g. Hambleton, 1993; Lawless, 1996). None of this theory is concerned with very general questions about 'the nature of planning'; rather it has focused on the specific problems of urban decay and regeneration in seeking to develop a better understanding of both the nature of the problems and the effectiveness of existing or possible future policies.

Social inequalities and equal opportunities

Although, in the 1980s and 1990s, unemployment and its attendant economic hardship has hit middle class people, implicit in the concern with inner city decline and regeneration has been an ongoing concern with the plight of the poorest and most disadvantaged groups in society. And this in turn has been associated with an ongoing concern with social inequalities in Britain.

The rhetoric of the New Right also included talk of a 'classless' society in which each person would have the opportunity to compete with others. This implied a commitment to equal opportunities for all, but if individuals could do better than others through competition, equal opportunities could result in inequalities of rewards. But the commitment to a classless society in which there was fair competition implied a commitment to equal opportunities in the first place, and in fact this commitment to equality of opportunity had long been a central plank of liberal political philosophy alongside its promulgation of the freedom of individuals within a market society.

To the political Left, the 'problem of inequality' has traditionally been viewed in terms of people's class position (e.g. Marx and Engels, 1848; Westergaard and Resler, 1975; Miliband, 1977). However, over the last two decades there has been a shift in the way such issues are conceived, to a perspective which incorporates other dimensions or sources of inequality and disadvantage, such as race, gender, disability, sexual orientation, age and so on. Thus some feminists have argued that gender is as significant as social class as a source of unequal treatment (e.g. Bryson, 1992, Chaps. 10–13).[8]

This is not the place to explore or evaluate the validity of this shift in thinking on the nature of social inequality and disadvantage, and the concommitant shift in thinking about equal opportunities. But we note two main points. First, that (as with the issue of economic decline and regeneration) a number of planning theorists have come to focus on developing theory about the specific and related issues of social inequality and planning for equal opportunities, as distinct from theory about the nature of planning as a whole. And second, the shift in thinking about these issues described above has been reflected in the planning theory literature. Thus, through the 1980s and 1990s research on social disadvantage and equal opportunities in relation to planning has come to focus predominantly on such matters as race, gender and disability rather than on inequalities attributable generally to social class.[9]

The global ecological crisis and sustainable development

In the late 1960s and early 1970s a wave of serious concern arose about the ecological damage being done to the natural environment by human activity (e.g. Carson's *Silent Spring*, the Club of Rome's *Limits to Growth* and Goldsmith *et al.*'s *A Blueprint for Survival*). This concern was also marked by the formation of a number of 'Green' political parties around the world (the British Ecology Party – subsequently renamed the Green Party – was formed in 1973). This first wave of ecological thinking filtered through into some planning theory. For example, the systems view of planning (see Chapter 4) was partly influenced by the idea of viewing the natural world as an 'ecosystem' (or a collection of inter-related ecosystems), and evaluating human actions in terms of their effects on the functioning of the natural ecosystem. Thus the opening chapter of the first British textbook advancing a systems view of planning (McLoughlin, 1969) was an account of 'Man in his ecological setting'. In this, Mcloughlin described some of the effects of mankind's actions on natural ecosystems as a way of illustrating the fact that we inhabit 'systems', that our actions affect the state of these systems, and that therefore it is vital to understand the likely effects of our actions on nature if we are to avoid environmental disasters.

For McLoughlin, ecological thinking was a model, and an object lesson, of systems thinking and its relevance to environmental planning. In spite of this, the effects of town planning actions on the environment were not a major concern of planning theorists throughout the 1970s. However, in the 1980s a number of environmental disasters (such as the accident at Chernobyl), together with some alarming scientific evidence of changes to the earth's global ecology (such as global warming and ozone depletion), combined to generate a second wave of environmental concern. The seriousness of these problems, together with their global reach, implied the need for international as well as national action, and in 1992 an 'Earth Summit' conference was held at Rio de Janeiro at which the world's nations gathered to try to agree some common action to avert environmental catastrophe.

Given the global scale of these problems, it might seem there is little town planning can do to help address them. Yet with its remit to plan and regulate the pattern and scale of new development, town planning has traditionally played some role in conserving certain environments and habitats; indeed, this was regarded by the Thatcher governments of the 1980s as the main achievement of Britain's post-war town planning system, and thus one of the main reasons for maintaining it (Heseltine, 1979). At the global scale, member states at the Rio Summit committed themselves to an action plan called 'Agenda 21', part of which was a key role for local governments ('local Agenda 21s'). The nature, location and density of urban land uses and the way people travel within and between cities, affect the 'ecology of the city' and, thus, global ecology. In the 1980s and 1990s there was a spate of theoretical work on planning for environmentally 'sustainable' cities (e.g. Breheny, 1992; Blowers, 1993; EU Expert Group, 1994; Barton, Davis and Guise, 1995; Buckingham-Hatfield and Evans, 1996). One issue of particular relevance to town planning is the question of whether the 'compact city' might, in general, be an environmentally more sustainable form than the dispersed cities which have arisen in the twentieth century (e.g. Jenks, Burton and Williams, 1996).

It was concern over environmental policy that largely prompted a switch in Conservative government strategy towards a more proactive, interventionist position on environmental planning. This was marked by the appointment of Chris Patten as Secretary of State for the Environment in 1988, and the subsequent publication of the white paper, *This Common Inheritance* (DoE, 1990).[10] The government also commissioned research into how best to appraise the 'environmental content' of development plans (DoE, 1993).

It is not my purpose here to summarise or discuss this developing body of theory. My main purpose is to draw attention to it to illustrate how some theorists have turned away from very general questions about 'the nature of planning', and focused instead on developing theory about a specific environmental problem and what planning might contribute to alleviating it. To call this a 'specific environmental problem' may understate the significance of this work for, clearly, if major components of the earth's natural ecology are irrecoverably altered, then Fukuyama's (1989) idea of the 'end of history' might become a grimmer, more literal reality. Understandably, those planning theorists working in the field of environmental sustainability regard their project as not just one amongst several branches of planning theory but the most fundamental kind of planning theory there is.

The aesthetic quality of urban environments

Planning theorists in the 1970s and early 1980s showed little interest in questions of urban design and aesthetics; indeed, some writers even suggested that attending to aesthetics was 'trivial' compared with, say, the income distributive effects of planning (Simmie and Hale, 1978). Town planning had become to many a 'social science' not an art. It is not surprising, therefore that 1980s research on the quality of design control exercised by British local planning authorities found that most planners were poorly qualified to exercise informed and sensitive control over the design quality of development proposals (Beer, 1983; Booth, 1983).

Yet this marginalisation of aesthetics and urban design in town planning changed in the 1980s. The roots of this change can be traced back to the 1960s when a number of architectural theorists turned against the anonymity of modernist 'functional' architecture and argued instead for the creation of a stylistically and therefore aesthetically richer architecture (e.g. Venturi, 1966). Out of this critique emerged 'postmodern' architecture, which self-consciously sought to bring back 'style' (i.e. a richer aesthetic content) to architecture (see Chapter 9). The debate about style was brought to wider public attention in 1984 when Prince Charles was invited to address the Royal Institute of British Architects. The Prince used the occasion to launch an attack on modern architecture in general, and the proposal for the extension to the National Gallery in particular, which he famously described as like 'a carbuncle on the face of a dearly loved friend'. Prince Charles subsequently presented his views on a BBC television programme and later in his book, *A Vision of Britain* (Prince of Wales, 1989). All this attracted public attention and triggered debate in both the architectural and town planning press (e.g. Hutchinson, 1989; Punter, 1990b). Although the focus of this debate was on architecture, it was bound to broaden out to the aesthetic quality of the built environment in general; indeed, Prince Charles's own view of architecture emphasised the importance of the landscape or townscape setting to the design of buildings.

Two further factors focused attention on aesthetics. First, one aspect of Conservative policy was to relax town planning's aesthetic control over new development (see DoE, 1980, para. 19). This *laissez-faire* approach attracted criticism (see, e.g. Hillman, 1990, especially the Foreword by Lord St John of Fawsley) and in *Planning Policy Guidance Note 1*, published at the beginning of the 1990s, the government withdrew its earlier advice in an annex which restated the importance of aesthetic considerations in town planning control (DoE, 1992a). Second, the President of the RTPI who has arguably had the biggest impact in that role since 1945 was the late Francis Tibbalds, who was President in 1988. And Tibbalds's main mission whilst President was to raise the status of urban design within the planning profession so that it was again seen as a central part of town planning.

By the beginning of the 1990s, therefore, the subject of aesthetics had once again come to the fore (see, e.g. Punter 1986; 1987; 1990a; 1994; Taylor, 1994a). Moreover, the government itself commissioned research into design policy in local planning (DoE, 1994a; 1994b).

'Cultural policy', which is broadly connected with the 'aesthetics of the city', also attracted some theoretical attention at this time. Urban cultural policy is primarily concerned with fostering the arts and cultural activities in cities. Recently it has come to be seen as having an important role in revitalising and regenerating central city areas which have suffered from the loss or decentralisation of traditional industries (such as ports) or other city centre activities (such as shops) (Bianchini and Parkinson, 1993). Cultural policy therefore ties in with the economic and social concerns described earlier.

The idea of 'culture' has often been a fairly traditional and restricted one, referring to 'the arts', and especially the established 'high' arts of concert music, opera, dance, plays, painting, etc. Much cultural policy has concentrated on the creation of 'cultural districts' with concert halls, theatres, art galleries, museums and so on. However, cultural policy has now widened to embrace a broader 'more democratic' conception of culture which includes all kinds of everyday social and leisure activities that take place in cities and that contribute to their economic and social vitality – activities such as walking, meeting others in cafés and pubs, eating out, sport, even work (Griffiths, 1993, p. 44). It is here that cultural policy connects with aesthetics and design, notably with the design of the public spaces within which the cultural life of the city takes place (see, e.g. Madanipour, 1996; Oc and Tiesdell, 1997).

Local democratic control and 'popular' planning

In streamlining planning procedures and in introducing measures to facilitate market-led development, the Thatcher governments bypassed or reduced opportunities for local democratic control over land development. In some areas this aroused local protest movements not unlike those of the 1960s. In London's docklands, for example, inhabitants of the Borough of Newham drew up a 'People's plan for the Royal Docks' in protest against the plans for the Royal Docks devised by the unelected and locally unaccountable London Docklands Development Corporation (Newham Docklands Forum/Greater London Council, 1984). These community planning movements were described as examples of 'popular planning' because they went beyond mere protest and involved communities formulating their own plans for their localities. In London, the Royal Docks popular plan proved to be ineffective against the muscle of the LDDC but other episodes of popular planning met with greater success, (e.g. the development of Coin Street, London) where a local community group succeeded in getting their own planning scheme implemented over that of a major developer (see Brindley, Rydin and Stoker, 1989). The 1980s also saw the rise of more militant environmental protests against the destruction of natural habitats and landscapes (e.g. the protests against motorway developments across Twyford Down or around Newbury).

These popular planning and protest movements ensured that questions of democracy and public participation were kept on the agenda and the subject of popular planning, and democratic planning more generally, became a further focus of attention for some planning theorists (see, e.g. Montgomery and Thornley, 1990, Chaps. 12–15).

PLANNING THEORY NOW

The foregoing discussion brings this account of urban planning theory since 1945 up to date. As this chapter has shown, at the beginning of the 1980s the whole idea of public sector town planning came under threat from the neo-liberal ideology of the New Right, but in the event, the state's role in town planning has remained, albeit in a changed political context. In fact, in the current climate of environmental concern, the perceived need for publicly accountable forms of environmental control and regulation has, if anything, grown stronger. For example, in Britain there remains a widespread acceptance of the need to control the scale of new 'edge city' developments which the free market strategy of Thatcherism encouraged, to plan carefully where new housing development should be and, above all, to plan development which is environmentally sustainable. In other words, the ongoing scale and seriousness of environmental problems has checked the forward march of free market liberalism and renewed the case for planning.

As shown in the last section, many planning theorists have turned their backs on 'grand theorising' about planning as a whole, and focused instead on the environmental problems which planning addresses. But – as the development of 'communicative planning theory' shows (see Chapter 7) – theoretical reflection on the nature of town planning in general has not been completely cast aside. We still have 'planning theory' in its wider (and traditional) sense, although clearly the *kinds* of general planning theories prevalent in the 1990s are very different from those which dominated town planning thought 50 years ago.

Some have said that, since the 1970s, planning theory has fragmented into a plurality of diverse theories (see, e.g. Healey, McDougall and Thomas, 1982a, Chap. 2; Hague, 1991, p. 300). Yet in one respect a distinction made a quarter of a century ago by Andreas Faludi (1973b, Chap. 1) continues to live on in contemporary planning theory, and to this extent there has been some continuity in the development of planning theory. Faludi (*ibid.*) distinguished between two types of theory: on the one hand, there was 'substantive' planning theory, which was about the object (towns, cities, the environment) town planning deals with including theory which aims to improve our understanding of the problems planning addresses; on the other there is 'procedural' planning theory, which is about the process of planning itself. Thus problem-centred research and theory described earlier is essentially substantive research and theory. By contrast, in viewing planning as a process of communication and deliberation, communicative planning theory is essentially procedural.[11]

Some planning theorists have suggested that procedural planning theory – the rational process view of planning – has been superseded by later theoretical developments (see Healey, McDougall and Thomas, 1982a, Chap. 2). However, communicative planning theory is also a procedural planning theory and central to the development of this theory has been the Habermasian ideal of a process of deliberation and decision-making which is 'undistorted'. This ideal can be regarded as a model of ideally rational planning and so, in this respect, communicative planning theory can be viewed as a further development of the rational process view of planning which developed over thirty years ago.

NOTES

1. As Roger Scruton (1980, Chap. 1) noted, liberalism was in many ways at odds with traditional conservatism. However, the New Right drew on conservative as well as liberal ideas; see Hall and Jacques (1983).
2. Critics of social democracy sometimes described their target as 'socialism' because of the high degree of state intervention and the important role of the public sector endorsed by social democrats. But in anything like the pure meaning of the term, social democracy was not socialism. For although social democrats advocated bringing some sectors of the economy into public ownership, they did not advocate the wholescale social ownership or control of the means of production. So if post-war social democracy was socialism, it was the 'revisionist' socialism promulgated, for example, by Anthony Crosland (1956). On the other hand, nor was social democracy synonymous with traditional free market liberalism because social democrats argued for a much greater degree of state intervention in civil society, and hence a much greater role for the public sector than classical liberals allowed. Again, if social democracy was a form of liberalism, it was revisionist or 'welfare' liberalism; it was the liberalism of Keynes (1946) and Rawls (1972), not of Friedmann (1962) and Nozick (1974).
3. A possible exception here is Jane Jacobs (1961), whose critique of town planning in both theory and practice comes close to implying that the free market would do better. For example, she cites cases of run-down city neighbourhoods which, if left alone from the clutches of town planners, would 'unslum' by themselves (Chap. 15).
4. In fact, Sorenson and Day's proposals for a 'libertarian' (or liberal) approach to planning are based on a close reading of Hayek's earlier proposals, which they largely endorse. Their biggest departure from Hayek is over the treatment of betterment, which they argue should not be taxed away but rather viewed as 'good fortune' if a result of public investment (e.g. in infrastructure), or a reward for 'entrepreneurial effort' if a result of legitimate market competition (Sorenson and Day, 1981, pp. 395–6). They also resist Hayek's blanket application of compensation payments for negative externalities (worsenment), arguing that these should not apply in cases where commercial competition results in losses to other property

owners and traders, for such losses are an inevitable consequence of healthy market competition. Like Hayek, Sorenson and Day seem to see little role for town planners in shaping the future pattern or form of urban land use and development by the production of development plans, though they concede that there might be a case for 'some notional land-use zoning' to reduce 'developer uncertainty' and 'neighbourhood unease' (*ibid.*, p. 393).

5. With regard to assessing the overall significance of the changes made to the planning system under New Right/Thatcherite ideology in the 1980s, a crucial point concerns the nature of the British planning system *prior* to the coming to power of Thatcherism. In many respects, and especially following the dismantling of the financial provisions for the taxing of 'betterment' which had been attached to the Town and Country Planning Act 1947, the British planning system had limited powers to produce development outcomes different from the market. Hence the view of those political economists, whose views I described in chapter 6, that land development was primarily determined by market forces, subject only to relatively minor constraining regulation by the planning system. In this respect, the planning system which the Conservative government found in place when it came to power in 1979 was *already* a market supportive one, and thus one which the government, with its explicit avowal of free market principles, had no need to alter fundamentally to serve its purposes. For fuller discussions of the significance of the changes to the British planning system under Thatcherism, see Thornley (1991, pp. 217–19 who argues that the changes were generally significant) and, for an opposing view, Griffiths (1990, cited in Montgomery and Thornley, 1990, pp. 21–33).
6. In this context, what is called 'planning gain' became an increasingly used practice by British planning authorities in the 1980s. Here, planning authorities negotiate 'Section 106' (formerly Section 52) agreements with private sector developers, as well as voluntary and public sector partners, to achieve certain development outcomes.
7. Cf. Stoker and Mossberger (1994, p. 195).
8. This shift in perspective mirrors the decline in the fortunes of socialism in the 1980s, and with it the class-based social theory which was a cornerstone of traditional socialist political theory.
9. See, e.g. Montgomery and Thornley (1990, Chaps. 16–19) on planning for equal opportunities generally; Thomas and Krishnarayan (1994) on planning and race; Little (1994) on planning and gender; Hahn (1986), Bennett (1988), and Imrie (1996) on planning and disability.
10. And further measures followed, in the form of new circular advice. Thus *Planning Policy Guidance Note* 12 advised local planning authorities to undertake environmental appraisals of their development plans (DoE, 1992b), whilst *PPG*13, on transport, stressed the importance of integrating land use and transport planning, and limiting the environmental damage caused by road traffic (DoE, 1994a).
11. Because of this it has been subjected to similar criticisms as were once levelled at the rational process model of planning of the 1960s and 1970s (see, e.g. Fainstein, 1995).

PART IV
CONCLUSION

9

Paradigm shifts, modernism and postmodernism

CHANGES IN PLANNING THOUGHT AND PARADIGMS

It will be clear from this book that ideas about town planning have changed over the fifty year period since the end of the Second World War. But what have been the most significant changes? In this concluding chapter I offer a retrospective overview of the evolution of town planning thought since 1945, and in doing so, I shall try to describe the most significant shifts in planning thought over this period.

At various times since 1945 overviews of post-war planning theory have appeared (e.g. Friedmann and Hudson, 1974; Galloway and Mahayni, 1977; Hemmens, 1980; Healey, McDougall and Thomas, 1982a; Yiftachel, 1989; Hague, 1991; Healey, 1991). Galloway and Mahayni (1977) speak of 'paradigm change' in post-war planning theory, and this term has come to be used widely by social theorists to describe major shifts in the history of ideas. It may therefore be useful to say something about the idea of 'paradigms' to begin with, and how this notion might relate to the recent history of town planning theory.

As used to describe changes in thought, the idea of a 'paradigm' derives from the work of Thomas Kuhn (1962), who employed the concept to describe major shifts in theoretical perspective in the history of science. According to Kuhn, if we look at the history of science, we find that advances in scientific thought have rarely occurred in a steady, evolutionary manner in response to the gradual accumulation of empirical evidence. Rather, the history of science is marked by long periods in which a given theoretical perspective – or 'paradigm' – has prevailed and been accepted by members of a scientific community. During these relatively stable periods, most scientific research is premissed upon the prevailing paradigm, and empirical observations are interpreted in terms of it. However, there is often some empirical evidence which does not 'fit' neatly with the prevailing theoretical claims. Many scientists seem willing to 'turn a blind eye' to this evidence on the assumption that one day someone will explain how it fits within the framework of the current paradigm. However, truly creative scientists are those who develop a new theoretical framework that succeeds in accounting for the hitherto puzzling evidence *as well as* the evidence previously explained by the 'old' paradigm. When a new paradigm succeeds in replacing an old paradigm in this way, there is a revolution in scientific thought. For a whole way of perceiving and explaining some aspect of the world is overthrown and replaced by a new theoretical perspective. As an example, the change from viewing the Earth as flat and at the centre of the Universe, to seeing it as round and as orbiting the Sun, was obviously a profound and revolutionary change in scientific thought – a paradigm shift in Kuhn's terms. Another example given by Kuhn is the shift in the 20th century from the Newtonian to Einsteinian view of space and time (Kuhn, 1962, Chap. 7). Once a new paradigm has become accepted, most scientific research comes to operate within this theoretical framework and, typically, another quiescent period of developing and refining this recently established theoretical framework ensues. So, according to Kuhn, the history of science continues.

It is clear from this account that, for Kuhn, paradigm changes are fundamental shifts in people's view of the world; that is why he describes paradigm shifts as revolutionary. And because they represent such fundamental changes in world view, such paradigmatic shifts typically occur infrequently in the history of science. Any given paradigm, once established, shapes the whole way a scientific community (and beyond that, the general public) views some aspect of the world, and tends to endure for centuries, not just decades.

We should therefore be cautious in applying the notion of paradigms, and paradigm shifts, to the changes in town planning thought which have occurred over the relatively short period of fifty years described in this book. We also need to note that Kuhn was describing changes in *scientific* thought; that is, major changes in the way people have *described and explained* some aspect of reality as a matter of fact. Town planning is not, in the strict sense, a science (not even – as some still persist in saying – a social science). Rather, it is a form of social action, directed at shaping the physical environment, and driven in this by certain moral, political, and aesthetic values. In other words, town planning is an 'ethical' (and hence political) practice, though of course, in seeking to realise certain valued ends, town planning should draw on relevant scientific understanding.

Nevertheless, providing we are alert to the dangers of overusing Kuhn's concept of paradigm shifts there is nothing to stop us using it in a looser or more generous way to describe significant changes in town planning thought.[1] Moreover, the notion of paradigm shifts may also be applied to fundamental shifts in values or ethical thinking.[2] With these precautions in mind, I shall explore the appropriateness of applying the Kuhnian terminology of paradigm shifts to describe the main changes in town planning theory since 1945.

This overview of post-war planning theory is organised as follows. In the next section I shall offer a view of the two most significant changes in town planning thought since 1945. The first of these occurred in the 1960s, with the shift from the urban design tradition to the systems and rational process views. The second change occurred during the 1970s and 1980s, and represented a shift in view of the planner's role. In particular, there was a shift from a view of the planner as a technical expert to the view of the planner as a kind of 'facilitator', drawing in other people's views and skills to the business of making planning judgements.

I then evaluate post-war planning thought from a more general perspective, situating changes in planning thought within the larger context of general changes in contemporary thought and culture. In particular, I describe the thesis put forward by some writers that from the late 1960s to the 1980s, a significant change in the history of western thought occurred from what has been called 'modernism' to 'postmodernism'. Arguably, this more general, cultural change in thought and values has also had a significant impact on town planning thought.

I conclude the chapter with some final reflections on town planning as a discipline, and the kind of theory that is relevant to this discipline.

TWO PARADIGM SHIFTS IN PLANNING THOUGHT SINCE 1945?

From town planning as design to science

Chapter 1 described how, for almost 20 years following the Second World War, British town planning theory and practice was dominated by a concept which saw town planning essentially as an exercise in physical *design*. Its long historical lineage is shown by the fact that, for as far back as we can see, what came to be seen as town planning was assumed to be most appropriately carried out by architects. Indeed, such was the intimate connection between architecture and town planning that the two were not distinguished throughout most of human history. Thus what we call town planning was seen *as* architecture; its only distinctiveness being that it was architecture on the larger scale of a whole town, or at least part of a town, as distinct from an individual building.

The concept of town planning as 'architecture writ large' persisted to the 1960s, as is shown by the fact that most planners in the post-war years were architects by training, or 'architect-planners'. Like architecture, town planning was viewed as an 'art', albeit (again like architecture) an 'applied' or 'practical' art in which utilitarian or 'functional' requirements had to be accommodated. Hence the systems and rational process views of planning which burst on to the scene in the 1960s represented a rupture with tradition – a change in planning thought which can be seen as a paradigm shift in the most fundamental, Kuhnian sense.

In Chapter 4 I stressed that the systems and rational process views of planning were conceptually distinct and thus really two theories of planning. Thus the systems view was based on a view about the *object* that town planning deals with (the town, or the environment in general, was viewed as a 'system'), whereas the rational process view concerned the *process* of planning itself. But both views, taken together, represented a departure from the prevailing design-based view of town planning. This shift in planning thought can be summarised under four points.

First, an essentially physical or morphological view of towns was replaced with a view of towns as systems of inter-related activities in a constant state of flux. Secondly, whereas town planners had viewed and judged towns predominently in physical and aesthetic terms, they now examined them in terms of social life and economic activities; in Harvey's (1973) terms, a sociological conception of space replaced a geographical or morphological conception of space. Thirdly, because the town was now seen as a 'live' functioning thing, this implied a 'process' rather than an 'end-state' or 'blueprint' approach. Fourthly, all these changes implied in turn a change in the kinds of skills appropriate to town planning. If town planners were trying to control and plan complex, dynamic systems, what seemed to be required were rigorous 'scientific' methods of analysis. Overall, the shift in planning thought suggested that town planning was a science, not an art. For the analysis of environmental systems involved systematic empirical investigation, while the concept of planning as a process of rational decision-making seemed to make planning a 'scientific' exercise because 'rationality' was equated with science.[3] This shift was so significant that it was profoundly unsettling to many planners and planning students reared in the design tradition of town planning. Suddenly, town planners who had seen themselves as 'artistic' urban designers were being told by a new generation of planning theorists that their former conception of town planning was inappropriate and that they should see themselves as 'scientific' systems analysts.

However it is important to note that the design-based tradition of town planning was not completely superseded by the theoretical changes of the 1960s. Although questions of design and aesthetics were marginalised in planning theory for over 20 years, in practice the development control sections of local planning authorities continued to assess development proposals in terms of their design and aesthetic impact. Moreover, the well-known *Essex Design Guide* (Essex County Council Planning Department, 1973), was the first sign that many local authorities were seeking to place their practice of design control on a clearer theoretical footing by articulating 'principles' of good design. This was significant. For it drew attention to the fact that, at the level of 'local' planning, the physical form and aesthetic appearance of new development remained a necessary and significant consideration for town planning. It was therefore only at the broader, more strategic level of planning that the design-based view of planning was supplanted by the changing conception of planning ushered in by the systems and rational process views of planning.

The revolution in town planning thought of the 1960s did not involve the *complete* replacement of one view of town planning by another. The real revolution was in making a distinction between two levels of town planning, one strategic and longer term, and the other 'local' and more immediate. The altered concept of town planning brought about by systems and rational process thinking was most appropriate at the *strategic* scale, although there were also lessons for 'local' small area planning in systems and rational process thinking (e.g. in giving greater consideration to social and economic factors, in approaching local planning as a rational process, etc.). In retrospect, then, the revolution in town planning thought in the 1960s was not a wholescale revolution which completely ousted the incumbent design-based view of town planning. Seen like this the advent of the systems and rational process views of planning did not represent a paradigm shift of the most fundamental kind described by Kuhn.

From the town planner as technical expert to 'communicator'

Although the shift from seeing town planning as an exercise in physical planning and design to seeing it as a rational process of decision-making directed at the analysis and control of urban systems was undoubtably a radical shift in thinking, there was one thing that the design-based view, and the systems and rational process views, had in common. It was that the town planner was someone who possessed some specialist skill – some expertise – which the average person in the street did not possess. It was this which qualified the planner, literally, to plan. And it was this, too, which made town planning a distinct 'profession' (the possession of some specialist knowledge or skill being the hallmark of any claim to professional status).

With such changes to planning thought there were inevitably changes in views about the specialist skills a planner required. The traditional design-based view of town planning demanded skills of aesthetic appreciation and urban design but the systems and rational process views demanded those of scientific and logical analysis. However, in the 1960s it was also acknowledged that town planning judgements were at root judgements of value (as distinct from purely technical judgements) about the kinds of environments it is desireable to create or conserve. The question therefore arose as to whether town planners had any greater 'specialist' ability to make these judgements than the ordinary person in the street. The experience of much 1960s planning – such as comprehensive housing redevelopment or urban road planning – seemed to indicate not. The view that town planning was a value-laden, political process therefore raised not so much the question of what the town planner's area of specialist expertise should be but, more fundamentally, the question of whether there was any such expertise at all.

From this questioning developed a curious bifurcation in planning theory which has persisted to this day. On the one hand, some planning theorists have continued to believe that the practice of town planning requires some specialist substantive knowledge or skills – be it about urban design, urban regeneration, sustainable development or whatever.[4] On the other has developed a tradition of planning thought which openly acknowledges that town planning judgements are value-laden and political. One conclusion would be to reject entirely the idea that town planning involves some special expertise at all, and some 'radical' planning theorists have flirted with this view (e.g. Goodman, 1972; Evans, 1995).[5] However, most planning theorists who have openly acknowledged the value-laden and political nature of planning have developed an alternative line of thought which rejects the idea that the town planner is someone who is specially qualified to make better decisions. What is 'better' is a matter of value, and planners have no superior expertise in making value-judgements about environmental options. However, the view is still taken that the town planner possesses some specialist skill, namely, skill in managing the process of arriving at planning decisions. A tradition of planning theory has emerged, therefore, which views the town planner's role as one of identifying and mediating between different interest groups involved in land development. The town planner is seen as someone who acts as a kind of cypher for other people's assessments of planning issues, rather than someone who is better qualified to assess these issues him or herself. The town planner is viewed as not so much a technical expert (i.e. as someone who possesses some superior skill to plan towns), but more as a 'facilitator' of other people's views about how a town, or part of a town, should be planned.

An early version of this theory was Paul Davidoff's 'advocacy' view of planning (see Chapter 5). The most recent version is the communicative planning theory described in Chapter 7, under which the skills of interpersonal communication and negotiation are seen to be central to a non-coercive, 'facilitator' model of town planning. It has even been suggested that the kinds of interpersonal skills needed by the communicative town planner are those of the listener and the counsellor:

> Meaningful dialogue – learning the language of the client – is at the heart of effective counselling. To counsel is not to give advice or push the client down a particular path, but to let the client see himself or herself fully and through this discovery achieve personal growth. As local government offices look for ways of including citizens in decision-making, they must adopt many counselling skills – active listening, non-judgemental acceptance, and the ability to empathise. How can people play a part in the decision-making process unless we 'enable' them to do so? (Healey and Gilroy, 1990, p. 22).

This is a far cry from the view that the specialist skill of the town planner resides in being either an urban designer or a systems analyst.

However, once again a word of caution is in order before it is too readily assumed that the difference between the two theories represents something like a paradigm shift, for it is possible to imagine some kind of merger between the two views. Thus a view of the town planner as primarily a communicator and negotiator can accommodate the planner having specialist knowledge which, for example, would enable him or her to point out the likely consequences of development proposals on the form and functioning of a town. Such a model of the town planner would be akin to that of, say, the civil servant who is an expert in economic matters, and who imparts his specialist economic understanding to those he advises who make decisions about economic policy. To be effective as an adviser, such a town planner would have to be skilled in communicating and negotiating with others, but he would also have to possess some specialist knowledge to bring to the communicating table to assist others in arriving at planning decisions.

MODERNISM AND POSTMODERNISM

According to some contemporary thinkers, over the last twenty years or so there has been a sea change in western thought and culture from 'modernism' to 'postmodernism' – a change so significant as to represent something akin to a Kuhnian paradigm shift. If there has been such a fundamental shift then town planning thought could hardly have escaped its impact. The alleged shift from modernism to postmodernism is particularly relevant to town planning as town planning and architecture have been one of the main 'sites' where the shift from modernism to postmodernism has most clearly occurred. Thus according to one of the leading writers about postmodernism, Charles Jencks (1991, p. 23), the modern era ended when, on 15 July 1972, the Pruitt-Igoe housing estate in St Louis (USA) – which had earlier won an award as an exemplar of modern architecture and town planning – was dynamited and destroyed by the local city authority.

Historical change is, of course, rarely as dramatic as this. The high-rise housing estates built all over Europe and North America have long been fiercely criticised as soulless, inhumane environments. The 'functional' style of modern architecture was criticised even in its infancy for its anonymity and lack of visual interest (e.g. the 1920s debate between the Dutch de Stijl architects, who favoured the austere geometrical forms which were to become the norm of modern architecture, and the adherents to the Amsterdam School of architecture, who favoured an architecture of expressive forms and decorative brickwork). However, from the late 1960s onwards the rejection of modernism became stronger and more widespread and developed into the movement now called postmodernism.

Postmodernism as a style

At one level, postmodernism represented a reaction against the styles of art and design which had been promoted by the modern movement. Postmodernists rejected the pared-down simplicity of modern 'functional' architecture, and so sought to 'bring back style' to enrich the aesthetic content of contemporary buildings. Thus Robert Venturi (1966, p. 16), in what is arguably the first text of postmodern architecture, famously counterposed his preference for a stylistically more complex architecture over plain 'functional' modernism:

> I like complexity and contradiction in architecture . . . Architects can no longer afford to be intimidated by the puritanically moral language of orthodox Modern architecture. I like elements which are hybrid rather than 'pure', compromising rather than 'clean', distorted rather than 'straightforward', ambiguous rather than 'articulated', . . . inconsistent and equivocal rather than direct and clear. I am for messy vitality over obvious unity . . . I am for richness of meaning rather than clarity of meaning.

A similar preference for complexity and 'richness' over simplicity and order was voiced by Jane Jacobs for town planning in *The Death and Life of Great American Cities* (1961). Jacobs berated modern city planning for its simple-mindedness, as represented, for example, in the single-use zoning of urban areas, or in its uncompromising approach to 'comprehensive' redevelopment which showed little understanding of the delicate social and economic fabric of so-called slum areas. Jacobs suggested that successful city areas were those with mixed uses, and that currently run-down 'slum' areas could, if left alone by planners, naturally 'unslum' themselves.

According to some accounts, however, the shift from modernism to postmodernism goes deeper than aesthetic preference. Underpinning the 'modern' movement generally was a more fundamental intellectual orientation involving a reliance upon reason and science – an optimistic belief that, through rational analysis and greater scientific understanding, humans could create a better world for themselves. The technological urban Utopias produced by modern architects in the early years of the twentieth century were expressions of this modernist faith in science and technology. Such was the self-confident optimism of architects like Le Corbusier and the Italian Futurists that they advocated sweeping away the traditional city to make way for their modern urban Utopias. Sixty or so years later, following the horrors wrought by modern technology, including the horrors of the Corbusier experiment of the comprehensively planned high-rise city, this confidence in modern science and technology has been seriously dented. Some postmodernists have offered a critique of the modernist reliance on science and even reason itself (e.g. Rorty, 1989; see also Truett Anderson, 1995). From this viewpoint, the dynamiting of the Pruitt-Igoe housing estate was not just an act to get rid of one housing form but it was also a symbol of a more fundamental collapse of the modernist faith in human rationality and scientific technology as the keys to a brighter human future.

Postmodernism as a challenge to science and reason

The modernist faith in reason and science had its roots in the European Enlightenment of the eighteenth century. What Habermas (1981) has called the 'project' of modernity was really a technological development of the Enlightenment, and the idea that cities could be made better by rational thought and action – by 'planning' – was thus part of this project. The postmodern critique therefore brings into question the Enlightenment assumption that the deficiencies of cities and our environment generally can be made better by planned action. Hence the real radicalness of Jacobs's critique was not just her preference for mixed uses over single use zoning, etc., but her implicit questioning of whether cities could be made better places by rational planning at all. This proposition is certainly a radical one, and again might suggest a paradigm shift in the most fundamental sense.

Some of the most radical versions of postmodernism therefore bring into question the efficacy of reason itself. Such a position is stated, for example, by Michael Dear (1995, p. 28, cited in Healey *et al.*, 1995):

> Postmodernism's principal target has been the rationality of the modern movement, especially its foundational character, its search for universal truth . . . The postmodern position is that all meta-narratives are suspect; that the authority claimed by any single explanation is ill-founded, and hence should be resisted. In essence, postmodernists assert that the relative merit of one meta-narrative over another is ultimately undecidable; and by extension, that any such attempts to forge intellectual consensus should be resisted.

Taken at face value this statement implies a rejection of rational discourse altogether. For if postmodernists believe, as Dear here suggests, that there are literally *no* criteria against which we can judge the relative merits of different theoretical positions, then it follows that there can be no *reasoned* debate about different theories at all. However, apart from the fact that such a position is intellectually hopeless (in the literal sense that there would be no point in hoping for greater enlightenment through rational discourse with others), it is also self-defeating. For if there are no rational criteria against which to judge the truth of a proposition, then there are no reasonable grounds for accepting as true the postmodern proposition that the 'relative merit of one meta-narrative over another is ultimately undecidable'. As Anthony Giddens (1990, p. 47) has written: 'Were anyone to hold such a view (and indeed if it is not inchoate in the first place), they could scarcely write a book about it. The only possibility would be to repudiate intellectual activity altogether.'

In fact, many writers who advance an extreme 'postmodernist' position of epistemological relativism are themselves inconsistent in adhering to it. Thus within a page of the position quoted above, we find Michael Dear writing about what the city of Los Angeles is 'really like' on the assumption, presumably, that there are truths about the place which can be discovered. The idea, then, that there are no standards of rationality which we should aspire towards in engaging in theoretical work should be dismissed, as Giddens (*ibid*., p. 46) again puts it, 'as unworthy of serious intellectual consideration'.

The same applies to those who criticise the ideal of rationality in relation to town planning. By all means, we may reject as undesirable and unrealistic the pursuit of *comprehensiveness* in town planning, whether in relation to the actual practice of replanning cities comprehensively or in relation to the process of making planning decisions. As argued earlier in this book (Chapter 4), the pursuit of rationality is distinct from the question of whether we are required to be comprehensive. But the proposition that rationality does not matter in planning theory, or that a rational approach to the process of planning is no better than any other approach, is open to the same criticisms as the postmodernist rejection of rationality made above.

If these arguments are persuasive, then the idea that postmodernism represents a paradigmatic break with Enlightenment reason turns out to be empty. The employment of reason in planning, aided by the best possible scientific understanding of the world we are seeking to plan, remains as relevant and important now as it has ever been. As for the critique of 'planning', it is important to recognise that what postmodernists have criticised is *modernist* planning (i.e. 'clean-sweep' comprehensive planning), rather than necessarily planning as such. After all, there can be different styles of town planning, some of which are compatible with the ideals articulated by Jacobs. The idea that some cities have grown 'naturally', as Alexander (1965) puts it, is misleading. For most human action is planned to some degree. We might talk of some cities having grown in more piecemeal, incremental and 'organic' ways, but then it is possible to envisage styles of town planning which are precisely like this.

Postmodernism as an alternative normative view of the quality of life

The conclusion reached in the previous section does not necessarily imply a rejection of postmodernism entirely, for postmodernism also represents a shift in thinking about style and aesthetics, and it is here that the real significance of postmodernism lies. Postmodernism, however, does not just question certain *styles* but posits some alternative *values* of a more general kind – an alternative view, in fact, of the quality of life. These alternative values bring into question many of the values and normative principles which have informed town planning in the modern age, and it is from this point of view that postmodernism presents a case for the serious reconsideration of the purposes of town planning.

What, then, are these alternative postmodern values? Postmodernists argue that the world and our experience of it is far more complex and subtle than has typically been realised. Thus in relation to cities, postmodernists claim that people's experience of places, and from this the qualities of places, are much more diverse and 'open' than was implicit in many modern schemes, and especially the bombastic simplicities of modern architectural visions of the ideal city. Instead of the modernist emphasis on simplicity, order, uniformity and tidiness, postmodernists typically celebrate complexity, diversity, difference, and pluralism (cf. Marion Young, 1990). Hence there can be no one type of environment which is ideal for everyone, no singular conception of environmental quality. Thus some may hold as an ideal Howard's genteel garden cities, but others will prefer the buzz and excitement of big city life – the 'teeming metropolis' as Elizabeth Wilson calls it (1991, p. 101).

Central to postmodernist values is a celebration of big city life because of its diversity and pluralism, and for the freedom of choice that this diversity promises. These values connect postmodernism with liberalism, for liberals also celebrate the plural society in which individuals have the opportunity to determine and 'realise' themselves through the exercise of free choice. It is these values which allow us to see Venturi (in relation to architecture) and Jacobs (in relation to town planning) as early postmodernists, for they argued for complexity and diversity as opposed to the sterile simplifications of modernism.

All this is very general and, indeed, one of the criticisms which can be levelled at the normative position of postmodernism is that it is so general as to be elusive. The celebration of diversity, for example, can be taken to an extreme where anything that is different may be accepted or permitted; in other words, a position of moral and political relativism corresponding to, and equally as untenable as, the epistemological relativism of some postmodernists discussed earlier. Thus in town planning, might there not be some values and ideals which town planning should aspire to, wherever it is practised? For example, shouldn't town planning, everywhere, do what it can to help bring about economically and environmentally sustainable development, development which is not socially divisive, and development which is experienced as an aesthetic delight? (In other words, shouldn't town planning be broadly guided by the ideals which underpin the five areas of 'problem-based' research and theory described in Chapter 8?) If so, then whatever postmodernists about the virtues of diversity and pluralism, and however important this is as a lesson for town planning, there may still be some overarching 'universal' ideals which town planning should aspire to.

CONCLUSION: TOWN PLANNING AS A DISCIPLINE AND TOWN PLANNING THEORY

In conclusion, we come back to town planning theory. At various times since 1945 town planning theorists have written about planning theory as if there could be, or should be, only one kind of 'planning theory'. Faludi's (1973b) early view that it was only theory about the planning process (procedural planning theory) which warranted the title 'planning theory' was an example of this. We also see this intellectual imperialism in the later rejection of procedural planning theory, and the alternative suggestion that town planning theory should necessarily be empirically based theory about the role of planning within its political economic context (e.g. Scott and Roweis, 1977).

The truth is that there are different *types* or *kinds* of theories, answering different kinds of questions, and not only one type of theory is relevant to town planning. Scientific theories which seek to improve our understanding of the world that town planning is dealing with, including the effects of different kinds of planned actions on the world, are fundamental to sound town planning. But town planning exists to improve the world, not just to understand it. Therefore philosophical reflection on the purposes of planning, such as that which postmodernism has prompted, is also central to planning theory. In other words, normative theory – including moral and political philosophy – is also a proper part of town planning theory.

Any account of what theory is relevant to town planning presupposes some conception of what sort of an activity town planning is, and much more could be said about this than space here allows. However, two broad observations about the nature of town planning further illuminate the kind of theory most relevant to its practice.

First, town planning is a form of *social action*, or a *social practice*. It is about intervening in the world to protect or change it in some way – to make it other than it would otherwise be without planning. Because it is a practice, it requires, more than anything, sound *judgement* – judgement about what best to do. In this respect, theory about practical reasoning and judgement is absolutely central to town planning, as John Forester has insisted in developing his version of communicative planning theory. Seen thus, town planning is neither an art nor a science in the strict sense of either of these terms, though, of course, sound judgement in town planning draws on both aesthetic and scientific understanding.

Because town planning is a practical discipline, some town planners have been sceptical about the value of town planning theory, and in Britain this has generated a most unfortunate 'anti-intellectualism' in the planning profession, as Reade (1987) has emphasised. Certainly, if a theory has no bearing on the planning, then there is no need for practising planners to that is not an argument against theory, only an argument it theory. And so, even if the rejection of certain kinds of tified, this does not justify the rejection of theory entirely. On ecisely because town planning is a practical discipline which the environment in which people live, it is all the more essential od practical theory to inform it. And, as we saw from our analysis of early post-war planning theory, what some people think of as practical 'common sense' is often not good practical theory at all.

The second main point to note is that town planning involves making judgements about what *best* to do – that is, about how best to plan the environments we inhabit. Throughout this book attention has been drawn to the fact that town planning is fundamentally about making value-judgements about the kinds of environments we want to protect or create. It follows that, at the heart of town planning theory, there should be rigorous analysis of, and theories about, environmental quality: what constitutes it; what different views (if any) different groups take of it; what different sorts of qualities (e.g. economic, social, aesthetic, ecological) make up quality environments and what possible tensions can arise between these different components of environmental quality; and how good-quality environments have been created in the past, and how they are most likely to be created in the conditions in which we now find ourselves.

One final thought. Of the two areas of practical and normative theoretical inquiry described above, we have, perhaps, got further with the former than with the latter. That is, since 1945, we have learnt more about how best to go about the practical process of town planning than we have learnt about the kinds of environmental qualities town planning should be aiming at. For one thing which emerges from this account is that considerably more theoretical attention has been devoted to refining our conception of what kind of an activity town planning is, and from this how best it should be approached, than has been devoted to analysing the constituents of high-quality environments and how they might be realised. If this is true, then it suggests that one of the most important tasks facing town planning theorists now is the development of better theory about the environmental qualities which town planning practice should help bring into being.

NOTES

1. Kuhn himself used the concept of a paradigm in different senses, some more fundamental than others (see Kuhn's 1969 postscript to his original essay, published in the second edition of 1970, pp. 174–91).
2. E.g. we could describe the shift in moral thinking that was ushered in by the European Enlightenment of the eighteenth century – in which all individual human beings came to be viewed as morally significant and therefore as possessing certain human 'rights' – as a fundamental change, or 'paradigm shift', in ethical thought.

3. This equation between rationality and science was of course too crude, for it implied that 'art' was somehow non-rational in comparison with science because, e.g. it involved the imagination, creativity, the expression of feelings, etc., rather than rigorous analysis. But this is misleading. For artistic activity involves rational analysis, and correspondingly, original scientific work involves creativity and imagination – as twentieth-century theoretical physics vividly illustrates.
4. Amongst those who take this view are those who continue to take a highly 'technicalist' view of the skills needed for planning, reminiscent, in fact, of the 'technicism' of the early systems planning theorists. This tradition persists, for example, in many of the articles which appear in the journal *Planning and Environment B: Planning and Design*.
5. In addition to these references, one could add some of the literature of the neoliberal New Right – such as Jones (1982) – which argued that public sector town planning did not achieve better outcomes than the free play of market forces. From this point of view, the so-called expertise of town planners was really no expertise at all.

Bibliography and references

Abercrombie, P. 1933: *Town and Country Planning*, London, Oxford University Press (3rd edition 1959).

Alexander, C. 1965: A city is not a tree, *Architectural Forum*, Vol. 122, no. 1 (reproduced in Bell and Thywitt, *op. cit.*, pp. 401–26).

Allison, L. 1975: *Environmental Planning: A Political and Philosophical Analysis*, London, Allen & Unwin.

Althusser, L. 1965: *Reading Capital*, London, New Left Books (1970 edition).

Amin, A. (editor) 1994: *Post-Fordism: A Reader*, Oxford, Blackwell.

Archbishop of Canterbury's Commission on Urban Priority Areas 1985: *Faith in the City: A Call for Action by Church and Nation*, London, Church House.

Arnstein, S.R. 1969: A ladder of citizen participation, *Journal of the American Institute of Planners*, Vol. 35, July, pp. 216–24.

Ayer, A.J. 1936: *Language, Truth and Logic*, Harmondsworth, Penguin (1946 edition).

Bachrach, P. and Baratz, M. 1963: Decisions and nondecisions: an analytical framework, *American Political Science Review*, Vol. 57, pp. 641–51.

Backwell, J. and Dickens, P. 1978: Town planning, mass loyalty and the restructuring of capital: the origins of the 1947 planning legislation revisited. *Urban and Regional Studies Working Paper* No. 11, Brighton, University of Sussex.

Bains Report 1972: *The New Local Authorities – Management and Structure*, London, HMSO.

Ball, M. 1983: *Housing Policy and Economic Power: the Political Economy of Owner-Occupation*, London, Methuen.

Banfield, E.C. 1959: Ends and means in planning, *International Social Science Journal*, Vol. XI, no. 3. (reprinted in Faludi, *op. cit.*, pp. 139–49).

Banham, R. 1971: *Los Angeles: The Architecture of Four Ecologies*, London, Penguin.

Bardach, E. 1977: *The Implementation Game: What Happens after a Bill Becomes a Law*, Cambridge, Mass. and London, MIT Press.

Barrett, S. and Fudge, C. (editors) 1981: *Policy and Action: Essays on the Implementation of Public Policy*, London, Methuen.

Barlow Commission 1940 (Royal Commission on the Distribution of the Industrial Population) *Report* (Cmd 6153) London: HMSO.

Barton, H., Davis, G. and Guise, R. 1995: *Sustainable Settlements: A Guide for Planners, Designers and Developers*, London, Local Government Management Board.

Bassett, K. 1996: Partnerships, business elites and urban politics: new forms of governance in an English city? *Urban Studies*, Vol. 33, no. 3, pp. 539–55.

Beer, A.R. 1983: Development control and design quality. Part 2. Attitudes to design, *Town Planning Review*, Vol. 54, no. 4, pp. 383–404.

Bell, D. 1960: *The End of Ideology*, New York, The Free Press.

Bell, G. and Thywitt, T. (editors) 1972: *Human Identity in the Urban Environment*, Harmondsworth, Penguin.

Bennett, T. 1988: Planning for disabled access, *Planning Practice and Research*, no. 4, Spring, pp. 8–10.

Berman, M. 1982: *All That is Solid Melts into Air: The Experience of Modernity*, London, Verso.

Berry, B.J.L. 1967: *Geography of Market Centers and Retail Distribution*, Englewood Cliffs, NJ, Prentice-Hall.

Bianchini, F. and Parkinson, M. (editors) 1993: *Cultural Policy and Urban Regeneration: The West European Experience*, Manchester, Manchester University Press.

Blowers, A. (editor) 1993: *Planning for a Sustainable Environment*, London, Earthscan.

Boddy, M. 1982: Planning, landownership and the state, in Paris, *op. cit.*, Part 2, pp. 83–94.

Boddy, M. and Fudge, C. (editors) 1984: *Local Socialism?* London, Macmillan.

Booth, P. 1983: Development control and design quality. Part 1. Conditions: a useful way of controlling design? *Town Planning Review*, Vol. 54, no. 3, pp. 265–84.

Braybrooke, D. and Lindblom, C.E. 1963: *A Strategy of Decision: Policy Evaluation as a Social Process*, New York, The Free Press.

Breheny, M. (editor) 1992: *Sustainable Development and Urban Form*, London, Pion.

Briggs, A. 1962: *William Morris: Selected Writings and Designs*, Harmondsworth, Penguin.

Brindley, T., Rydin, Y. and Stoker, G. 1989: *Remaking Planning: The Politics of Urban Change in the Thatcher Years*, London, Unwin Hyman.

Broady, M. 1968: *Planning for People*, London, Bedford Square Press.

Brown, M. 1966: Urban form, *Journal of the Town Planning Institute*, Vol. 52, pp. 3–10.

Brown, M. 1968: The time element in planning, *Journal of the Town Planning Institute*, Vol. 54, pp. 209–13.

Bruton, M.J. (editor) 1974: *The Spirit and Purpose of Planning*, London, Hutchinson.

Bryson, V. 1992: *Feminist Political Theory: An Introduction*, London, Macmillan.

Buchanan, C.D. *et al.* 1963: *Traffic in Towns*, London, HMSO (shortened edition published by Penguin, Harmondsworth, 1964).

Buckingham-Hatfield, S. and Evans, B. 1996: *Environmental Planning and Sustainability*, Chichester, Wiley.

Burchell, R.W. and Sternlieb, G. (editors) 1978: *Planning Theory in the 1980s: A Search for Future Directions*, New Brunswick, NJ, Rutgers University Press.

Camhis, M. 1979: *Planning Theory and Philosophy*, London, Tavistock.

Campbell, S. and Fainstein, S. (editors) 1996: *Readings in Planning Theory*, Oxford, Blackwell.

Carson, R. 1962: *Silent Spring*, Harmondsworth: Penguin.

Castells, M. 1977: *The Urban Question: A Marxist Approach*, London, Edward Arnold (2nd edition, translated by Alan Sheridan).

Chadwick, G.F. 1966: A systems view of planning, *Journal of the Town Planning Institute*, Vol. 52, pp. 184–6.

Chadwick, G.F. 1971: *A Systems View of Planning*, Oxford, Pergamon Press.

Cherry, G.E. 1974: *The Evolution of British Town Planning*, London, Leonard Hill Books.

Clark, K. 1969: *Civilisation*, London, John Murray and the BBC.

Cockburn, C. 1977: *The Local State: Management of Cities and People*, London, Pluto Press.

Cohen, G.A. 1978: *Karl Marx's Theory of History: A Defence*, Oxford, Clarendon Press.

Cooke, P. (editor) 1989: *Localities: The Changing Face of Urban Britain*, London, Unwin Hyman.

Coser, L. 1956: *The Functions of Social Conflict*, New York, The Free Press.

Cox, A. 1984: *Adversary Politics and Land*, Cambridge, Cambridge University Press.

Creese, W.L. 1967: *The Legacy of Raymond Unwin: A Human Pattern for Planning*, Cambridge, Mass., MIT Press.

Crosland, C.A.R. 1956: *The Future of Socialism*, London: Jonathan Cape.

Cullingworth, J.B. 1975: *Environmental Planning: 1939–1969. Volume 1. Reconstruction and Land Use Planning, 1939–1947*, London, HMSO.

Dahrendorf, R. 1969: *Class and Class Conflict in Industrial Society*, London, Routledge & Kegan Paul.

Davidoff, P. 1965: Advocacy and pluralism in planning, *Journal of the American Institute of Planners*, Vol. 31, November (reprinted in Faludi, 1973a *op. cit.*, pp. 277–96.

Davidoff, P. and Reiner, T.A. 1962: A choice theory of planning, *Journal of the American Institute of Planners*, Vol. 28, May (reprinted in Faludi, 1973a, *op. cit.*, pp. 11–39.

Davies, J.G. 1972: *The Evangelistic Bureaucrat: A Study of a Planning Exercise in Newcastle-upon-Tyne*, London, Tavistock.

Dear, M. 1995: Prolegomena to a Postmodern Urbanism, In Healey *et al.*, 1995, *op. cit.*, Part 1, pp. 27–44.

Dear, M. and Scott, A.J. (editors) 1981: *Urbanisation and Urban Planning in Capitalist Society*, London, Methuen.

Dennis, N. 1970: *People and Planning: The Sociology of Housing in Sunderland*, London, Faber & Faber.

Dennis, N. 1972: *Public Participation and Planners' Blight*, London, Faber & Faber.

Department of the Environment 1969: *People and Planning: Report of the Committee on Public Participation in Planning* (the Skeffington report), London, HMSO.

Department of the Environment 1977: *Policy for the Inner Cities* (Cmnd 6845), London, HMSO.

Department of the Environment 1980: *Development Control – Policy and Practice* (Circular 22/80), London and Cardiff, HMSO.

Department of the Environment 1990: *This Common Inheritance* (Cm 1200), London, HMSO.

Department of the Environment 1992a: *Planning Policy Guidance Note 1: General Policy and Principles* (PPG1), London and Cardiff, HMSO.

Department of the Environment 1992b: *Planning Policy Guidance Note 12: Development Plans and Regional Planning Guidance* (PPG12), London, HMSO.

Department of the Environment 1993: *The Environmental Appraisal of Development Plans: A Good Practice Guide*, London, HMSO.

Department of the Environment 1994a: *Planning Policy Guidance Note 13: Transport*, London, HMSO.

Department of the Environment 1994b: *Design Policies in Local Plans: A Research Report*, London, HMSO.

Department of the Environment 1994c: *Design Policies in Local Plans: A Good Practice Guide*, London, HMSO.

DiGaetano, A. and Klemanski, J.S. 1993: Urban regimes in comparative perspective: the politics of urban development in Britain, *Urban Affairs Quarterly*, Vol. 29, no. 1, pp. 54–83.

Docherty, T. (editor) 1993: *Postmodernism: A Reader*, London, Harvester Wheatsheaf.
Drabble, M. 1991: A vision of the real city in Fisher, M. and Owen, U. (editors): *Whose Cities?* London, Penguin.
Duhl, L.J. (editor) 1963: *The Urban Condition: People and Policy in the Metropolis*, New York, Simon & Schuster.
Dunleavy, P. 1980: *Urban Political Analysis: The Politics of Collective Consumption*, London, Macmillan.
Dworkin, R. 1977: *Taking Rights Seriously*, London, Duckworth.
Esher, L. 1981: *A Broken Wave: The Rebuilding of England 1940–1980*, Harmondsworth, Penguin.
Essex County Council Planning Department 1973: *A Design Guide for Residential Areas*, Chelmsford, Essex County Council.
Etzioni, A. 1967: Mixed-scanning: a 'third' approach to decision-making', *Public Administration Review*, December (reprinted in Faludi, 1973a, *op. cit.*, pp. 217–29.
EU Expert Group on the Urban Environment 1994: *European Sustainable Cities*, Bristol, UWE.
Evans, B. 1995: *Experts and Environmental Planning*, Aldershot, Avebury.
Fainstein, N.I. and Fainstein, S.S. 1979: New debates in urban planning: the impact of Marxist theory within the United States, *International Journal of Urban and Regional Research*, Vol. 3, no. 3, September (reprinted in Paris, 1982 *op. cit.*, pp. 147–73.
Fainstein, S.S. 1995: Politics, economics, and planning: why urban regimes matter', *Planning Theory*, no. 14, pp. 34–41.
Fagence, M. 1977: *Citizen Participation in Planning*, Oxford, Pergamon Press.
Faludi, A. 1971: Problems with problem-solving, *Journal of the Royal Town Planning Institute*, November, p. 415.
Faludi, A. (editor) 1973a: *A Reader in Planning Theory*, Oxford, Pergamon Press .
Faludi, A. 1973b: *Planning Theory*, Oxford, Pergamon Press.
Faludi, A. 1979: Towards a combined paradigm of planning theory? *Planning Outlook*, Vol. 22, no. 2 (reprinted in Paris, 1982 *op. cit.*, pp. 27–38).
Faludi, A. 1987: *A Decision-Centred View of Environmental Planning*, Oxford, Pergamon Press.
Fischer, F. and Forester, J. 1993: *The Argumentative Turn in Policy Analysis and Planning*, London, University College Press.
Fisher, R. and Ury, W. 1981: *Getting to Yes: Negotiating Agreement Without Giving In*, London, Hutchinson.
Fishman, R. 1977: *Urban Utopias in the Twentieth Century: Ebenezer Howard, Frank Lloyd Wright, and Le Corbusier*, New York, Basic Books.
Foley, D. 1960: British town planning: one ideology or three?' *British Journal of Sociology*, Vol. 11 (reprinted in Faludi, 1973a, *op. cit.*, Part 2, pp. 69–93).
Forester, J. 1989: *Planning in the Face of Power*, Berkeley, Calif., University of California Press.
Friedmann, J. 1969: Notes on societal action, *Journal of the American Institute of Planners*, Vol. 35, pp. 311–18.
Friedmann, J. 1973: *Retracking America: A Theory of Transactive Planning*, New York, Anchor Press/Doubleday.
Friedmann, J. and Hudson, B. 1974: Knowledge and action: a guide to planning theory, *Journal of the American Institute of Planners*, Vol. 40, no. 1, pp. 2–16.
Friedman, M. 1962: *Capitalism and Freedom*, Chicago, Ill., University of Chicago Press.
Fukuyama, F. 1989: The end of history? *The National Interest*, Summer, pp. 3–18.
Galbraith, J.K. 1958: *The Affluent Society*, Harmondsworth, Penguin.

Galloway, T.G. and Mahayni, R.G. 1977: Planning theory in retrospect: the process of paradigm change, *Journal of the American Institute of Planners*, Vol. 43, no. 1, pp. 62–71.
Garnier, T. 1917: *Une Cité Industrielle*, Paris.
Geddes, P. 1915: *Cities in Evolution: An Introduction to the Town Planning Movement and the Study of Civics*, London, Ernest Benn (1968 edition).
Gibberd, F. 1953: *Town Design*, London, The Architectural Press.
Giddens, A. 1971: *Capitalism and Modern Social Theory: An Analysis of the Writings of Marx, Durkheim and Max Weber*, Cambridge, Cambridge University Press.
Giddens, A. 1990: *The Consequences of Modernity*, Oxford, Polity Press.
Giddens, A. 1994: *Beyond Left and Right: The Future of Radical Politics*, Oxford, Polity Press.
Glass, R. 1948: *The Social Background of a Plan*, London, Routledge & Kegan Paul.
Glass, R. 1959: The evaluation of planning: some sociological considerations', *International Social Science Journal*, Vol. Xl, no. 3 (reprinted in Faludi, 1973a, *op. cit.*, Part 2, pp. 45–67).
Goldsmith, E. Allen, R., Allaby, M., Davoll, J. and Lawrence, S. 1972: A blueprint for survival, *The Ecologist*, Vol. 2, no. 1, pp. 1–44.
Goodman, R. 1972: *After the Planners*, Harmondsworth, Penguin.
Gough, I. 1979: *The Political Economy of the Welfare State*, London, Macmillan.
Graham, K. 1986: *The Battle of Democracy*, Brighton, Harvester Wheatsheaf.
Griffiths, R. 1986: Planning in retreat? Town planning and the market in the eighties, *Planning Practice and Research*, no. 1, September (reprinted in Montgomery and Thornley, *op. cit.*, pp. 21–33).
Griffiths, R. 1993: The politics of cultural policy in urban regeneration strategies, *Policy and Politics*, Vol. 21, no. 1, pp. 39–46.
Gutch, R. 1972: The use of goals, *Journal of the Royal Town Planning Institute*, June, p. 264.
Gutkind, E.A. 1969: *Urban Development in Southern Europe: Italy and Greece* (Volume IV of Gutkind's *International Library of City Development*, London, Collier Macmillan.
Gyford, J. 1976: *Local Politics in Britain*, London, Croom Helm.
Gyford, J. 1985: *The Politics of Local Socialism*, London, Allen & Unwin, 1985.
Habermas, J. 1979: *Communication and the Evolution of Society*, London, Heinemann, (translated by T. McCarthy).
Habermas, J. 1981: Modernity – an incomplete project, *New German Critique*, Vol. 22, Winter (reprinted in Docherty *op. cit.*, Part 2, pp. 98–109).
Haggett, P. 1965: *Locational Analysis in Human Geography*, London, Arnold.
Hague, C. 1991: A review of planning theory in Britain, *Town Planning Review*, Vol. 62, no. 3, pp. 295–310.
Hahn, H. 1986: Disability and the urban environment: a perspective on Los Angeles, *Environment and Planning D: Society and Space*, Vol. 4 pp. 273–88.
Hall, P. 1974: The containment of urban England, *The Geographical Journal*, Vol. 4, Part 3, pp. 386–417.
Hall, P. 1975: *Urban and Regional Planning*, Harmondsworth, Penguin.
Hall, S. and Jacques, M. 1983 (eds.) *The Politics of Thatcherism*, London: Lawrence & Wishart.
Hall, P., Gracey, H., Drewett, R. and Thomas, R. 1973: *The Containment of Urban England*, London, Allen & Unwin.
Hambleton, R. 1993: Issues for urban policy in the 1990s, *Town Planning Review*, Vol. 64, no. 3, pp. 313–28.

Harloe, M. (editor) 1977: *Captive Cities: Studies in the Political Economy of Cities and Regions*, London, Wiley.
Harvey, D. 1973: *Social Justice and the City*, London, Arnold.
Harvey, D. 1989: From managerialism to entrepreneurialism: the transformation in urban governance in late capitalism, *Geografiska Annaler*, Vol. 71 B, pp. 3–17.
Hayek, F.A. 1944: *The Road to Serfdom*, London, Routledge & Kegan Paul.
Hayek, F.A. 1960: *The Constitution of Liberty*, London, Routledge & Kegan Paul.
Healey, P. 1991: Debates in planning thought, In Thomas and Healey *op. cit.*, pp. 11–33.
Healey, P. 1992a: A planner's day: knowledge and action in communicative practice, *Journal of the American Planning Association*, Vol. 58, Winter, pp. 9–20.
Healey, P. 1992b: Planning through debate: the communicative turn in planning theory, *Town Planning Review*, Vol. 63, no. 2, pp. 143–62.
Healey, P., Cameron, S., Davoudi, S., Graham, S. and Madani-Pour, A. (editors) 1995: *Managing Cities: The New Urban Context*, Chichester, Wiley.
Healey, P. and Gilroy, R. 1990: Towards a people-sensitive planning' *Planning Practice and Research*, Vol. 5, no. 2, pp. 21–9.
Healey, P., McDougall, G. and Thomas, M.J. (editors) 1982a: *Planning Theory: Prospects for the 1980s*, Oxford, Pergamon Press.
Healey, P., McDougall, G. and Thomas, M.J. 1982b: Theoretical debates in planning: towards a coherent dialogue, In Healey, McDougall and Thomas, *op. cit.*., pp. 5–22.
Held, D. 1987: *Models of Democracy*, Oxford, Polity Press.
Hemmens, G.C. 1980: New directions in planning theory, *Journal of the American Planning Association*, Vol. 46, no. 3, pp. 259–84.
Heseltine, M. 1979: Secretary of State's address, *Report of Proceedings of Town and Country Planning Summer School: 8–19th September 1979*, London, Royal Town Planning Institute, pp. 25–30.
Hill, M. 1968: A goals-achievements matrix for evaluating alternative plans, *Journal of the American Institute of Planners*, Vol. 34, pp. 19–29.
Hillman, J. 1990: *Planning for Beauty*, London, Royal Fine Arts Commission and HMSO.
Hobsbawm, E. 1994: *Age of Extremes: The Short Twentieth Century 1914–1991*, London, Abacus.
Howard, E. 1898: *Garden Cities of Tomorrow* (originally published as *Tomorrow: A Peaceful Path to Real Reform*), London, Faber & Faber, 1965.
Hutchinson, M. 1989: *The Prince of Wales: Right or Wrong?* London, Faber & Faber.
Imrie, R. 1996: *Disability and the City: International Perspectives*, London, Paul Chapman.
Imrie, R. and Thomas, H. 1993: *British Urban Policy and the Urban Development Corporations*, London, Paul Chapman.
Innes, J.E. 1995: Planning theory's emerging paradigm: communicative action and interactive practice, *Journal of Planning Education and Research*, Vol. 14, no. 3, pp. 183–9.
Jacobs, J. 1961: *The Death and Life of Great American Cities*, Harmondsworth, Penguin (1964 edition).
Jay, L.S. 1967: Scientific Method in Planning, *Journal of the Town Planning Institute*, Vol. 53, no. 1, pp. 3–6.
Jencks, C. 1991: *The Language of Post-Modern Architecture*, London, Academy Editions.
Jenks, M., Burton, E. and Williams, K. (editors) 1996: *The Compact City: A Sustainable Urban Form?* London, E & FN Spon.

Jones, R. 1982: *Town and Country Chaos: A Critical Analysis of Britain's Planning System*, London, The Adam Smith Institute.
Keeble, L. 1952: *Principles and Practice of Town and Country Planning*, London, The Estates Gazette.
Kemp, R. 1980: Planning, legitimation, and the development of nuclear energy: a critical theoretic analysis of the Windscale inquiry, *International Journal of Urban and Regional Research*, Vol. 4, no. 3, pp. 350–71.
Keynes, J.M. 1936: *The General Theory of Employment, Interest and Money*, London: Macmillan 1971 (Vol. VII of *The Collected Works of John Maynard Keynes*).
Kirk, G. 1980: *Urban Planning in a Capitalist Society*, London, Croom Helm.
Klosterman, R.E. 1978: Foundations for normative planning, *Journal of the American Institute of Planners*, Vol. 44, no. 1, pp. 37–46.
Kuhn, T. 1962: *The Structure of Scientific Revolutions*, Chicago, Ill., University of Chicago Press (2nd edition 1970).
Lawless, P. 1996: The inner cities: towards a new agenda', *Town Planning Review*, Vol. 67, no. 1, pp. 21–43.
Le Corbusier 1924: *Urbanisme*, Paris, Editions Cres (English translation by Frederick Etchells published as *The City of Tomorrow*, London, The Architectural Press, 1971).
Le Corbusier 1933: *La Ville Radieuse*, Paris, Vincent, Freal et Cie (English translation by Pamela Knight, Eleanor Levieux and Derek Coltman, published as *The Radiant City*, London, Faber & Faber, 1967).
Le Grand, J. and Estrin, S. (editors) 1989: *Market Socialism*, Oxford, Clarendon Press.
Lichfield, N. 1996: *Community Impact Evaluation*, London, UCL Press.
Lichfield, N., Kettle, P. and Whitbread, M. 1975: *Evaluation in the Planning Process*, Oxford, Pergamon Press.
Lindblom, C.E. 1959: The science of 'muddling through, *Public Administration Review*, Spring, (reprinted in Faludi, *op. cit.*, 1973a pp. 151–69).
Little, J. 1994: *Gender, Planning and the Policy Process*, Oxford, Pergamon Press.
Locke, J. 1690: *Two Treatises of Government*, London, Dent (1924 edition).
Long, N.E. 1959: Planning and politics in urban development, *Journal of the American Institute of Planners*, Vol. 25, no. 6, pp. 167–9.
Lukes, S. 1974: *Power: A Radical View*, London, Macmillan.
Macpherson, C.B. 1977: *The Life and Times of Liberal Democracy*, Oxford, Oxford University Press.
Madanipour, A. 1996: *Design of Urban Space: An Enquiry into a Socio-spatial Process*, Chichester, Wiley.
Magee, B. 1973: *Popper*, Glasgow, Collins/Fontana.
Mandelbaum, S.J., Mazza, L. and Burchell, R.W. (editors) 1996: *Explorations in Planning Theory*, New Brunswick, NJ, Rutgers University Press.
Marion Young, I. 1990: *Justice and the Politics of Difference*, Princeton, NJ, Princeton University Press.
Marx, K. 1859: *A Contribution to the Critique of Political Economy*, London, Lawrence & Wishart (1971 edition).
Marx, K. 1869: *The Eighteenth Brumaire of Louis Bonaparte* (reprinted in Marx and Engels, 1959, *op. cit.*).
Marx, K. and Engels, F. 1846: *The German Ideology*, London, Lawrence & Wishart (1974 edition edited by C.J. Arthur).
Marx, K. and Engels, F. 1848: *Manifesto of the Communist Party* (reprinted in Marx and Engels, 1959, *op. cit.*, pp. 43–82).
Marx, K. and Engels, F. 1959: *Basic Writings on Politics and Philosophy*, London, Collins Fontana (edited by L.S. Feuer).

McDougall, G. 1979: The state, capital and land: the history of town planning revisited, *International Journal of Urban and Regional Research*, Vol. 3, no. 3, pp. 361–80.

McKay, D.H. and Cox, A. 1979: *The Politics of Urban Change*, London, Croom Helm.

McLoughlin, J.B. 1965a: The planning profession: new directions, *Journal of the Town Planning Institute*, Vol. 51, pp. 258 61.

McLoughlin, J.B. 1965b: Notes on the nature of physical change – toward a view of physical planning, *Journal of the Town Planning Institute*, Vol. 51, pp. 397–400.

McLoughlin, J.B. 1969: *Urban and Regional Planning: A Systems Approach*, London, Faber & Faber.

Meadows, D.H., Meadows, D.L., Randers, J. and Behrens, W.W. 1972: *The Limits of Growth*, London: Earth Island.

Mellor, J.R. 1977: *Urban Sociology in an Urbanised Society*, London, Routledge & Kegan Paul.

Mellor, J.R. 1982: *Images of the City: Their Impact on British Urban Policy*, Open University Course D202: Urban Change and Conflict. Block 1, Unit 2, Milton Keynes, Open University Press.

Meyerson, M. and Banfield, E.C. 1955: *Politics, Planning and the Public Interest*, Glencoe Ill., The Free Press.

Miliband, R. 1969: *The State in Capitalist Society*, London, Quartet Books (1973 edition).

Miliband, R. 1977: *Marxism and Politics*, Oxford, Oxford Unversity Press.

Miller, D. 1989: *Market, State, and Community: Theoretical Foundations of Market Socialism*, Oxford, Clarendon Press.

Ministry of Housing and Local Government 1965: *The Future of Development Plans* (the Report of the Planning Advisory Group – the PAG Report), London, HMSO.

Montgomery, J. and Thornley, A. 1990: *Radical Planning Initiatives: New Directions for Urban Planning in the 1990s*, Aldershot, Gower.

More, T. 1516: *Utopia* (translated and introduced by Paul Turner), Harmondsworth, Penguin, 1965.

Morris, W. 1890: *News From Nowhere* (reprinted in Briggs, *op. cit.*).

Nairn, I. 1955: *Outrage*, London, Architectural Press.

Needham, B. 1971: Concrete problems, not abstract goals: planning as problem-solving, *Journal of the Royal Town Planning Institute*, July/August, pp. 317–19.

Newham Docklands Forum/Greater London Council 1984: *The People's Plan for the Royal Docks* London, Greater London Council.

Nove, A. 1983: *The Economics of Feasible Socialism*, London, Allen & Unwin.

Nozick, R. 1974: *Anarchy, State and Utopia*, Oxford, Blackwell.

Oatley, N. 1995: Editorial: urban regeneration, *Planning Practice and Research*, Vol. 10, nos. 3 and 4, pp. 261–9.

Oatley, N. *et al.* 1995: Urban regeneration, *Planning Practice and Research* (special issue), Vol. 10, nos. 3 and 4.

Oc, T. and Tiesdell, S. (editors) 1997: *Safer City Centres: Reviving the Public Realm*, London, Paul Chapman.

O'Connor, J. 1973: *The Fiscal Crisis of the State*, London, St James Press.

Pahl, R.E. 1970: *Whose City? And Further Essays on Urban Society*, Harmondsworth, Penguin, (2nd edition 1975).

Paris, C. (editor) 1982: *Critical Readings in Planning Theory*, Oxford, Pergamon Press.

Pateman, C. 1970: *Participation and Democratic Theory*, Cambridge, Cambridge University Press.

Perry, C.A. 1939: *Housing for the Machine Age*, New York, Russell Sage Foundation.

Pickvance, C. (editor) 1976: *Urban Sociology: Critical Essays*, London, Tavistock.

Pickvance, C. 1977: Physical planning and market forces in urban development, *National Westminster Bank Quarterly Review*, August (reprinted in slightly modified form in Paris, *op. cit.*, Part 2, pp. 69–82).
Popper, K.R. 1957: *The Poverty of Historicism*, London, Routledge & Kegan Paul.
Popper, K.R. 1963: *Conjectures and Refutations: The Growth of Scientific Knowledge*, London, Routledge & Kegan Paul.
Powell, E. 1969: *Freedom and Reality*, London, Elliot Right Way Books.
Pressman, J.L. and Wildavsky, A. 1973: *Implementation*, Berkeley, Calif., and London, University of California Press (3rd edition 1984).
Prince of Wales 1989: *A Vision of Britain: A Personal View of Architecture*, London, Doubleday.
Punter, J. 1986: A history of aesthetic control: Part 1. 1909–1953, *Town Planning Review*, Vol. 57, no. 4, pp. 351–81.
Punter, J. 1987: A history of aesthetic control. Part 2. 1953–1985, *Town Planning Review*, Vol. 58, no. 1, pp. 29–62.
Punter, J. 1990a: *Design Control in Bristol: 1940–1990*, Bristol, Redcliffe Press.
Punter, J. 1990b: The Ten Commandments of architecture and urban design' *The Planner*, 5 October, pp. 10–14.
Punter, J. (editor) 1994: Design research and planning practice, *Planning Practice and Research* (special issue), Vol. 9, no. 3.
Raiffa, H. 1982: *The Art and Science of Negotiation*, Cambridge, Mass. and London, Harvard University Press.
Ravetz, A. 1980: *Remaking Cities: Contradictions of the Recent Urban Environment*, London, Croom Helm.
Rawls, J. 1972: *A Theory of Justice*, Oxford, Oxford University Press.
Reade, E. 1987: *British Town and Country Planning* Milton Keynes, Open University Press.
Rees, G. and Lambert, J. 1985: *Cities in Crisis: The Political Economy of Urban Development in Post-War Britain*, London, Arnold.
Rex, J. 1970: *Key Problems of Sociological Theory*, London, Routledge & Kegan Paul.
Richards, J.M. 1950: New London office buildings: the Lessor scheme critically examined, *Architect's Journal*, Vol. 111, pp. 394–8.
Robinson, I.M. 1972: *Decision-Making in Urban Planning: An Introduction to New Methodologies*, Beverly Hills, Calif., and London, Sage Publications.
Rorig, F. 1932/1955: *The Medieval Town*, London, Batsford, 1967 (translated by B.T. Batsford).
Rorty, R. 1989: *Contingency, Irony, and Solidarity*, Cambridge, Cambridge University Press.
Sager, T. 1994: *Communicative Planning Theory*, Aldershot, Avebury.
Scott, A.J. and Roweis, S.T. 1977: Urban planning in theory and practice: a reappraisal, *Environment and Planning A*, Vol. 9, pp. 1097–119.
Scruton, R. 1980: *The Meaning of Conservatism*, second edition, London: Macmillan.
Sharp, T. 1940: *Town Planning*, Harmondsworth, Penguin.
Sharp, T. 1946: *Exeter Phoenix: A Plan for Rebuilding*, London, The Architectural Press.
Simmie, J.M. and Hale, D.J. 1978: The distributional effects of ownership and control of land use in Oxford, *Urban Studies*, Vol. 15, no. 1, pp. 9–21.
Simon, H.A. 1945: *Administrative Behaviour*, New York: Free Press.
Simon, H.A. 1960: *The New Science of Management Decision*, New York, Harper & Row.
Sked, A. and Cook, C. 1993: *Post-War Britain: A Political History*, Harmondsworth, Penguin (4th edition).

Soja, E. 1995: Postmodern urbanisation: the six restructurings of Los Angeles, in Watson and Gibson, *op. cit.*, pp. 125–37.

Sorenson, A.D. 1982: Planning comes of age: a liberal perspective, *The Planner (The Journal of the Royal Town Planning Institute)*, Vol. 68, no. 6, pp. 184–8.

Sorenson, A.D. 1983: Towards a market theory of planning, *The Planner (The Journal of the Royal Town Planning Institute)*, Vol. 69, no. 3, pp. 78–80.

Sorenson, A.D. and Day, R. 1981: Libertarian planning, *Town Planning Review*, Vol. 52, no. 4, pp. 390–402.

Stein, C.S. 1958: *Toward New Towns for America*, Liverpool, Liverpool University Press.

Stewart, M. 1987: Ten years of inner cities policy, *Town Planning Review*, Vol. 58, no. 2, pp. 129–45.

Stoker, G. and Mossberger, K. 1994: Urban regime theory in comparative perspective, *Environment and Planning C: Government and Policy*, Vol. 12, pp. 195–212.

Stone, C.L. 1989: *Regime Politics: Governing Atlanta 1946–1988*, Lawrence, Kans, University of Kansas Press.

Stone, C.N. 1993: Urban regimes and the capacity to govern: a political economy approach, *Journal of Urban Affairs*, Vol. 15, no. 1, pp. 1–28.

Susskind, L. and Cruikshank, J. 1987: *Breaking the Impasse: Consensual Approaches to Resolving Public Disputes*, New York, Basic Books.

Taylor, N. 1984: A critique of materialist critiques of procedural planning theory, *Environment and Planning B: Planning and Design*, Vol. 11, pp. 103–26.

Taylor, N. 1985: The usefulness of a conceptual theory of rational planning: a reply to Scott's comment, *Environment and Planning B: Planning and Design*, Vol. 12. pp. 235–40.

Taylor, N. 1994a: Aesthetic judgement and environmental design: is it entirely subjective?' *Town Planning Review*, Vol. 65, no. 1, pp. 21–40.

Taylor, N. 1994b: Environmental issues and the public interest, in Thomas, *op. cit.*, pp. 87–115.

Thomas, H. (editor) 1994: *Values and Planning*, Aldershot, Avebury.

Thomas, H. and Healey, P. (editors) 1991: *Dilemmas of Planning Practice*, Aldershot, Avebury.

Thomas, H. and Krishnarayan, V. (editors) 1994: *Race Equality and Planning: Policies and Procedures*, Aldershot, Avebury.

Thomas, M.J. 1979: The procedural planning theory of A. Faludi, *Planning Outlook*, Vol. 22, no. 2 (reprinted in Paris, *op. cit.*, pp. 13–25).

Thornley, A. 1991: *Urban Planning Under Thatcherism: The Challenge of the Market*, London, Routledge.

Tribe, L.H. 1972: Policy science: analysis or ideology? *Philosophy and Public Affairs*, Vol. 2, no. 1, pp. 66–110.

Tripp, H.A. 1942: *Town Planning and Road Traffic*, London, Arnold.

Truett Anderson, W. (editor) 1995: *The Fontana Postmodernism Reader*, London, Fontana.

Unwin, R. 1930: Regional planning with reference to Greater London (reprinted in Creese, *op. cit.*).

Uthwatt Committee 1942: *Final Report of the Expert Committee on Compensation and Betterment* (Cmnd 6386), London, HMSO.

Venturi, R. 1966: *Complexity and Contradiction in Architecture*, New York, Museum of Modern Art (2nd edition 1977).

Walzer, M. 1970: A day in the life of a socialist citizen, in Walzer, M. *Obligations*, Cambridge, Mass., Harvard University Press, pp. 229–38.

Ward, S.V. 1994: *Planning and Urban Change*, London, Paul Chapman.
Watson, S. and Gibson, K. (editors) 1995: *Postmodern Cities and Spaces*, Oxford, Blackwell.
Webber, M.M. 1963: The prospects for policies planning, in: Duhl, *op. cit.*, Part 4, pp. 319–30.
Webber, M.M. 1969: Planning in an environment of change. Part II. Permissive planning, *Town Planning Review*, Vol. XX, no. 1, pp. 277–95.
Westergaard, J. and Resler, H. 1975: *Class in a Capitalist Society: A Study of Contemporary Britain*, Harmondsworth, Penguin.
Williams, R. 1973: *The Country and the City*, London, Chatto & Windus.
Wilson, E. 1991: *The Sphinx in the City*, London, Virago.
Yiftachel, O. 1989: Towards a new typology of urban planning theories, *Environment and Planning B: Planning and Design*, Vol. 16, pp. 23–39.
Young, M. and Willmott, P. 1957: *Family and Kinship in East London*, Harmondsworth, Penguin (revised edition 1962).

Index

Abercrombie, Patrick 4, 8, 50
action (and social action) 111, 113–125
advocacy planning 85, 162
aesthetics x, 10–11, 23, 27–29, 30, 40, 146, 150–151, 159–160, 161
agenda 21 (and local agenda 21) 149
Alexander, Christopher 13–14, 33, 48–50, 55, 62, 97, 98, 109, 110, 165
Allison, Lincoln 80
Althusser, Louis 105
Amin, A 144
Amsterdam 8, 163
anti-urbanism 23, 27–29, 33, 47–48, 54
architecture 8, 40
architectural determinism 7, 42
Arnstein, Sherry 88–89
art 10, 65, 73, 159–160, 168
Ayer, Alfred J 90

Backwell, Jenny 106
Bains Report 69
Ball, Michael 107
Banfield, Edward 34, 50–51, 83
Banham, Reyner 36
Bardach, Eugene 121–122, 127
Barlow Report 18
Barrett, Susan (and Fudge, Colin) 119–121
Barton, Hugh 149
Bassett, Keith 145
Bauer, Catherine 47
Beer, Anne 150
Bell, Daniel 69, 77, 78, 132, 134
Bennett, Toni 154
Bentham, Jeremy 80
Berlage, HP 8
Berman, Marshall 76, 90
Berry, Brian 110
betterment 36, 134, 153, 154
Bianchini, Franco 151
Birmingham 141
Blowers, Andrew 149
blueprint plans 5, 14–17, 44–46, 52, 54, 69, 129, 160
Boddy, Martin 129, 143
Booth, Philip 150
Braybrooke, David 72
Breheny, Michael 149
Brindley, T. (and Rydin, Y. and Stoker, G.) 142–144, 151
Bristol 76, 90
Broady, Maurice 7, 42
Brown, Maurice 13, 44–45, 52
Bryson, Valerie 148
Buchanan, Colin 29–31, 37
Buchanan Report 29–31, 39, 65
Buckingham-Hatfield, S 149

Camhis, Marios 96, 110
Campbell, Scott viii
capitalism 104–107, 112, 126–129, 131, 132
Carson, Rachel 148
Castells, Manuel 105
Chadwick, George F 45, 61, 64, 74, 78, 90
Charles, Prince of Wales 150
Chandigargh 48, 50, 54
Cherry, Gordon F 8, 18, 27, 55
Chicago 83
Church of England 147
circulars/circular advice (UK government) 138–139
civil engineering 4, 8
Clark, Kenneth 37
class/class theory 105, 106, 148
Club of Rome 148
Cockburn, Cynthia 105
compulsory competitive tendering (CCT) 140
communication (interpersonal) 117, 122–125, 145, 160–162
communicative action (theory of) 122–125, 145, 153
communicative planning theory 122–125, 153, 167
community/communities 40–42
comprehensive planning, 23–27, 75, 163, 164, 165
conflict (over values, interests, etc) 50
consensus (over values, politics, etc) 21–22, 23, 34–35, 50–51, 54
conservatism 86, 130, 153
Conservative Party/Government 21, 129, 130, 136–139
consultation, public 43–44, 86–90
containment, urban 99–101
cost-benefit analysis 68, 79–81
counselling 162

Cox, Andrew 36, 129
Crosland, Anthony 153
Cullingworth, JB 21, 36
culture/cultural policy 139, 151
cybernetics 60, 65, 73

Dahrendorf, Ralph 50
Davidoff, Paul 84–85, 162
Davies, Jon Gower 51, 76, 83
Dear, Michael 105, 164–165
decision-making 66–69, 112, 113, 129
democracy 88–90, 123–125, 127–128, 146, 151–152
Dennis, Norman 43, 51, 76, 83
design (town planning as) 4, 5, 8–14, 159–160
development plans 27, 44–46, 51–54, 86–87
Dickens, Peter 106
Di Gaetano, A (and Klemanski, JS) 141
disjointed incremental planning 71–73, 95, 112–113, 129
Distribution of Industry Act 1945 18
Drabble, Margaret x
Dunleavy, Patrick 91

ecology/ecological thinking 65, 146
ecological crisis 148–150
economic decline/regeneration 108, 146–147
economics 159, 162
empirical (investigation, theory, etc) 96–109
Enterprise Zones 137
Environment (UK Department of) 138
environmental sustainability 146, 148–150, 161
entrepreneurialism 139–144
equal opportunities 146, 147–148
Esher, Lionel 39
Essex Design Guide 160
Etzioni, Amitai 72–73, 95
European Enlightenment 4, 73–74, 75, 164, 168
evaluation (of plans) 67–68, 78–81, 112
Evans, Bob 161
Exeter 11, 22

Fagence, Michael 86
Fainstein, Susan (and Fainstein, Norman) viii, 106, 144–145, 154
Faludi, Andreas viii, 61, 66, 69, 71, 72, 81–82, 83, 95, 96, 97, 112, 152, 167
Fischer, F 123
Fisher, R (and Ury, W) 122
Fishman, Robert 17, 23
flexibility (in planning) 52–53
Foley, Donald 6, 33, 34, 36
fordism 144
Forester, John 123, 125, 127, 167
France 17
Friedman, Milton 133, 153,
Friedmann, John 96, 113–115, 116, 117–118, 120, 122, 157
Fukuyama, Francis 134, 150
function (e.g. of architecture, the city, etc) 11–13, 46–47, 48–50, 61–64
Futurists (Italian) 164

Galbraith, John K 38
Galloway, TG 157
garden cities/garden city planning 17, 20–21, 23, 24, 29, 47–48
Garnier, Tony 8, 24
Geddes, Patrick 18, 29, 62, 66
geography 65
Gibberd, Frederick 4, 8
Giddens, Anthony 60, 73, 90, 104, 165
Glass, Ruth 14, 36, 42
Glasgow 143, 144, 145
goals/aims/objectives, etc (in general) 66, 67, 68, 78, 112
Goldsmith, Edward 148
Goodman, Robert 89, 161
Gough, Ian 105, 106
Gracey, Harry 28
grand theory 145, 152
Greater London Council 143, 144
Griffiths, Ron 151, 154
Gutch, Richard 112
Gutkind, EA 55
Gyford, Jon 86, 91, 143

Habermas, Jurgen 123–125, 164
Haggett, Peter 65
Hague, Cliff 152, 157
Hall, Peter 14, 28, 29, 74, 99–101, 110, 132
Hall, Stuart 153
Hambleton, Robin 147
Harloe, Michael 105
Harvey, David 105, 110, 140, 159
Hayek, Friedrich 132–133, 134–135, 153–154
Healey, Patsy 123, 152, 153, 157, 162
Held, David 90
Heseltine, Michael 136–139, 149
Hill, Morris 112, 129
Hillman, Judy 150
historical materialism 104
Hobsbawm, Eric 38
Howard, Ebenezer 17, 20–21, 23, 24, 29, 47–48
Hutchinson, Max 150

ideology 127
implementation 37, 111–122, 125–128, 145
Imrie, Rob 147, 154
inequalities (social) 100–101, 147–148
inner cities 40–44, 146–147
Innes, Judith 118, 123
Institute of Civil Engineers 18
instrumental (mean-end) reasoning 71

Jacobs, Jane 46–48, 50, 62, 65, 97, 98, 109, 110, 153, 163, 164, 165, 166
Jacques, Martin 153
Jay, LS 82
Jencks, Charles 163
Jones, Robert 134–135, 169
Joyce, James 36
judgement 167–168
justice (theory of) 98

Keeble, Lewis 4–13, 16–18, 24, 31, 34, 37, 59–60, 63, 64, 77, 95
Kemp, Ray 124
Keynes, JM/Keynsianism 21, 144, 153
Kirk, Gwyneth 105

Kuhn, Thomas 64, 157–158, 159, 160, 168

Labour Government 3, 21–22, 139, 146
ladder (of participation) 88–89
land 99–100
Lawless, Paul 147
Le Corbusier 8, 17, 18, 23, 24, 32, 36, 47, 48, 50, 54, 55, 74, 76, 164
Le Grand, Julian 133
liberalism 21, 90, 128–129, 130–134, 139, 168
Lichfield, Nathaniel 68, 79–81, 90, 112, 129
Lindblom, Charles 72, 112–113, 129
Little, Jo 154
local government/local authorities 139–144
local planning 52, 54, 160
local socialism 143
logical positivism 90
London 24, 40–41, 50, 76, 151
London County Council 24, 40
London Docklands Development Corporation (LDDC) 151
Long, Norton 83
Los Angeles 31, 36, 165
Lukes, Steven 110

Macmillan, Harold 38, 55
Madanipour, Ali 151
Magee, Bryan 82
Mahayni, RG 157
managerialism 39, 69, 101–102, 107
Marion Young, Iris 166
market socialism 133
markets/market forces 102–103, 104, 107, 110, 126, 131–135
Marx, Karl (and Engels, Frederick) 104, 110, 148
Marxism 101, 104–107, 126–127,
master plans/master planning 5, 14–17, 27, 44–46, 129
McDougall, Glen 106, 152, 153, 157
McKay, David 36, 107, 129
McLoughlin, JB 45, 59–60, 61–64, 65, 74, 90, 112, 129, 148
Mellor, Rosemary 28–29, 36
Meyerson, M 34, 50–51, 83
Middlesbrough 14, 42
Miliband, Ralph 105, 148
Miller, David 133
Milton Keynes 18
mixed land uses 48, 50
mixed scanning 72–73, 95
model village movement 20
modernism 22, 24–26, 36, 60, 73–74, 75–76, 90, 162–166
monitoring 68, 108
More, Thomas 23
Morris, William 20, 23
motorways 76
Mumford, Lewis 47, 55

Nairn, Ian 40
National Parks and Access to the Countryside Act 1949 21
Needham, Barrie 112
negotiation 117, 122–123, 145
neighbourhood effects 134
neighbourhoods/neighbourhood planning 31–33, 41–42
Netherlands 8, 17, 22, 163
new towns 7, 13, 18, 48, 50
New Towns Act 1946 21
New Right 129, 130–139, 146, 154, 168
Newcastle-upon-Tyne 51, 76
normative theory 3, 20–37, 70–71, 77–83, 129, 165–166, 167
Nove, Alec 133
Nozick, Robert 133, 153

Oakland (California) 115–116
Oc, Taner 151
O'Connor, James 133
Oatley, Nick 147
operations research 65
Owen, Robert 20

Pahl, Ray 39, 101–102
painting (modern) 90
paradigms/paradigm shifts 157–160
Paris 89
Patten, Chris 149
Perry, Clarence 32–33
philosophy xi, 75, 90
physical planning (town planning as) 4–8, 14, 17, 20, 39–42, 59–60, 159–160
physical determinism 7, 41–42, 54
Pickvance, Chris 39, 102–104, 107, 126
plan evaluation 67–68, 78–81,
Planning Advisory Group Report (PAG Report) 39, 51–54, 63, 86–87
planning gain 154
Planning Policy Guidance Notes (PPGs) 138
planning theory viii-ix, 152–153, 167–168
Plato 89
pluralism/pluralist theory 142, 166
policy and action 118–122
political economy 101–107, 125–129, 139–145
politics/political nature of planning 7, 35, 51, 77–90, 161
Popper, KR 69, 81–82, 109
popular planning 143, 151–152
positive planning 21
positivism (see also logical positivism) 90
post-fordism 144–145
post-modernism 162–166
Powell, Enoch 133
practical reasoning 126
Pressman, JL (and Wildavsky, A) 112, 115–117, 118, 122
problems (analysis of) 67–68
procedural planning theory (see also rational process view of planning) 60–61, 66–73, 95–98, 111–112, 145, 153
process view of planning 45, 60–61, 66–73, 159–160
profession (of town planning) 35, 84–85, 161
professions/professionalism 35, 84–85
protest (urban) 75–77, 89, 151–152
public interest 34, 128
public participation 77, 85–90, 123–125, 146
Punter, John 22, 150, 151

quantification 65

Radburn layouts 37
Raiffa, H 122
rationality (in general) 70, 164–165, 168
rational comprehensive planning 73–74, 95, 112–113
rational process view of planning 60–61, 66–74, 76, 77–78, 81, 95–98, 111–112, 153, 159–160, 161
Ravetz, Alison 25–27
Rawls, John 80, 98, 153
Reade, Eric 14, 27, 108–109, 128, 167
Reaganism 133
Rees, Gareth (and Lambert, John) 105
regime theory 131, 139–144
regional planning 18, 28
regulation 107
regulation theory 131, 144–145
Reiner, Thomas 84
Reith, Lord 34
relativism 166
representative democracy 86–90
Restriction of Ribbon Development Act 1935 28
Rex, Jonathan 50
Richards, JM 40
Rio de Janeiro (Summit Conference) 149
Rorig, F 55
Rorty, Richard 164
Royal Institute of British Architects (RIBA) 18
Royal Institute of Chartered Surveyors (RICS) 18
Royal Town Planning Institute (RTPI) 8, 18, 136

Sager, Toni 123
Sant'Elia, Antonio 24, 25
science 65, 69, 73–74, 81–83, 159–160, 164–165, 167, 168
scientific method 69, 81–83
Scott, Allen 96, 105, 106, 110, 167
Scruton, Roger 153
Sharp, Thomas 4, 8, 10–11, 22
Sheffield 143
Silkin, Lewis 43
Simon, Herbert 69
Simmie, James 150
Skeffington Report 77, 87–88, 89–90
social (including social planning) 6, 40–41, 54, 146, 159
social democracy/social democratic consensus 3, 7, 21–22, 69, 77, 130, 131–134, 144, 153
social engineering 109
socialism 20–22, 131, 132, 133, 139, 143, 144, 153, 154
Soja, Edward 36
Sorenson, Anthony (and Day, Richard) 135, 153–154
Soria y Mata 16
St John of Fawsley, Lord 150
Stein, Clarence 37, 47
Stewart, Murray 146, 147
Stoker, G (and Mossberger, K) 141, 154
Stone, C 140–142
strategic planning 18, 52–54, 160
structure planning 52, 54, 63
suburbanisation 99–100
Sunderland 43, 51, 54, 76
'survey-analysis-plan' 66–67
Susskind, L (and Cruikshank, J) 122
sustainability 148–149
systems view of planning 59–61, 61–66, 73–74, 76, 77–78, 81, 148–149, 159–160, 161

Taylor, Nigel 34, 55, 110, 151
technical view of planning 34–35, 43, 77–78
Thatcher, Margaret 129, 130, 135, 136, 137, 141, 144, 149
Thatcherism 130, 133, 136–139, 146, 152, 154
Thomas, Huw 147, 154
Thomas, MJ 96, 97, 110, 152, 154
Thornley, Andrew 137, 154
Tibbalds, Francis 150
Town and Country Planning Acts (UK):
1909 106
1947 4, 14–15, 18, 19, 21–22, 36, 44, 77, 86, 87, 102, 106, 132, 134, 154
1968 52, 63, 64, 87, 102
1990 138
Town Planning Institute (see also Royal Town Planning Institute) 8, 18, 45
town plans 14–17, 27, 44–46,51–54
traffic/transport planning 29–31, 39, 65
Tribe, Laurence 81–83
Tripp, Alker 37
Truett Anderson, Walter 164

Unwin, Raymond 10, 29
urban containment 28, 30, 99–101
urban design (including town planning as) 5, 8–14, 16–17, 39–40, 41–42, 53–54, 63–64, 146, 150–151, 159–160, 161, 163–164
Urban Development Corporations (UDCs) 137
urban regeneration 146, 161
urban sprawl 28
urban structure 23, 48–50, 54
Use Classes Order (UK) 137
Uthwatt Committee (Report of) 21–22, 36
utilitarianism 79–81
Utopianism (including 'Utopian comprehensiveness') 16–17, 20, 23–27, 46–48, 54, 75, 109, 164

values (in general) 83–85, 161, 165–166
values (of British town planning) 20–37, 43, 76, 166
Venturi, Robert 150, 163, 166
villages/urban villages 33

Ward, Stephen 21, 22
Webber, Melvin 51, 66
welfare state 4, 131
Westergaard, John 148
Williams, Raymond ix, 36, 37
Wilson, Elizabeth 29, 36, 166
Wordsworth, William 36
Wright, Frank Lloyd 17, 30, 36
Wright, Henry 37, 47

Yiftachel, O 157
Young, Peter (and Willmott, Michael) 40–41, 132

zoning (land use) 14–16, 31–32, 33–34, 44–45, 163, 164.